STONE
An introduction

STONE

An introduction

Second edition

ASHER SHADMON

Practical Action Publishing Ltd
27a Albert Street, Rugby, CV21 2SG, Warwickshire, UK
www.practicalactionpublishing.org

First published 1996\Digitised 2008
Reprinted in 2018

ISBN 10: 1 85339 313 4
ISBN 13: 9781853393136
ISBN Library Ebook: 9781780446042
Book DOI: http://dx.doi.org/10.3362/9781780446042

A catalogue record for this book is available from the British Library.

Since 1974, Practical Action Publishing has published and disseminated
books and information in support of international development work
throughout the world. Practical Action Publishing is a trading name
of Practical Action Publishing Ltd (Company Reg. No. 1159018), the
wholly owned publishing company of Practical Action. Practical Action
Publishing trades only in support of its parent charity objectives and any
profits are covenanted back to Practical Action (Charity Reg. No. 247257,
Group VAT Registration No. 880 9924 76).

Typeset by J&L Composition Ltd, Filey, North Yorkshire, UK

Contents

Tables and illustrations

Acknowledgements

Text drawings by Trevor Ridley, Dick Inglis and Ethan Danielson.

The author and the publishers would like to thank the following for their permission to reproduce the photographs on the pages noted; and apologize if they have not been able to trace all owners.

National Film Board of Canada (p. xii); Y. Barzilay, Press photographer, Shuafat, P.O. Box 19/0428, Ramallah Road, Jerusalem (p. 8); Perrier Fernand (p. 55); Gardner Denver Co (p. 72); Geological Survey of Britain (p. 93); Ricarda Schwerin, Jerusalem (p. 118); Israel Department of Antiquities and Museums (p. 123 top); United Nations/Rothstein (p. 123 bottom); Milton Grant (UN) (p. 141, 148 top); Knudsen-Moore Inc. (p. 156); BBC Hulton Picture Library (p. 157).

Foreword to second edition

Since the first edition was completed in 1987, significant technological developments have changed the working and application of stone. The volume of stone turnover has increased considerably and new technologies have made it possible to adapt stone to all kinds of shapes in urban furnishings, an application which now constitutes 10 per cent of the market.

An initial objective of this introduction to stone was to tie together traditional and modern methods and terminologies, and the book may well be a last link between the past and the future. Stone use is becoming widespread and the distinction between modest dry-walling and a sophisticated stone facade becomes simplified by sharing the same basic knowledge of stone properties and technologies required in both applications.

As in geology, where the present is the key to the past, traditional methods are the roots of intermediate technologies from which our advanced technical know-how and *savoir-faire* is sprouting. Unfortunately, training, especially of architects, is lagging behind the increased use of stone, leaving future contenders for refurbishment and restoration which are often as costly as, or more costly than, the original applications.

Important developments in technology since the mid 1980s include the use of diamond wire in stone sawing at the quarries, with a significant environmental bonus in facilitating quarry rehabilitation. Research into waste disposal and conversion has advanced considerably, as well as reduction in dust and noise. However, as before, the basic tools required are few and durable, and a habitable shelter can even be constructed without tools from field-stones or rocks of the right shape.

Stonehenge

Geotechnical knowledge has not kept pace with the new sources and varieties of stone, and there is an urgent need to rectify this if major failures of the kind which have occurred in Chicago and Helsinki are to be avoided. A reconsideration of standard tests and specifications, some of which date back to the early part of the twentieth century, would also be a positive step.

At the forefront of the quest for a better and safer environment are the problems of slipperiness of pavings and floors and the soiling of facades, the application of urban furnishings, and the much neglected rehabilitation and replication of dimension stone quarries.

In tackling these and other problems, consultants specializing in all aspects of stone can take an increasing role in ensuring its future development.

Asher Shadmon

Foreword to first edition

Imagine a building material that is available all over the world: that is durable, easy to use and versatile; a material that can be worked with tools found in any workshop to make simple structures, or used with high-technology machinery to produce lavish architectural constructions. Some new wonder material? A new-fangled product of the synthetic society? Far from it; this material has been in use for thousands of years. It is, of course, stone.

Yet so low is the building industry's regard for stone that it has been dropped from the syllabuses of many architectural courses and the agendas of international meetings. This book on dimension stone aims to put stone back on the map and to show why it is an important building material.

The contents cover many aspects of stone: what it is, where to find it, how to identify different stone types, extraction, architectural uses, industrial production and the development of stone industries. Examples of famous historical constructions bear witness to the 'durability' of stone: if the Tower of Babel had been built from stone, rather than bricks, it might still exist today, at least in relics!

It is hoped that this book will contribute to a wider use of stone in developing and industrialized countries alike; that it will encourage the development of small industries, especially in the former. It will be most useful to those who already have some knowledge of stone; it is, however, dedicated to the uninitiated, who may be living in an area where stone is readily available, yet who have never realized its potential.

When I was in Fiji, I visited a tribe reputed to be amongst the greatest wood-carvers in the Pacific. When I suggested that they try using stone as a carving material, they replied that they had not got the necessary tools. I asked them to show me the tools they used for wood-carving and they produced a mallet and a chisel. I picked up a piece of volcanic rock and carved a crude face. Everyone was amazed – they had no idea that stone could be worked so easily, and with woodworking tools, to boot. I suspect that many others have felt the same. Perhaps this book will show the way.

Asher Shadmon

It would be difficult to find material as versatile as stone. Above: Sea-wall built from marble rubble, Philippines. Below left: Windmill made from limestone.
Below right: Example of an inukshook, a type of megalith found in the northernmost parts of America.

1 Nature's own building material

Early use

Stone is undoubtedly the oldest construction material known to man. It existed before animal life evolved and plant life vegetated: indeed, the history of stone preceded the history of man. Stone was the material used in man's earliest dwellings: natural caves were followed by caves hewn out of rock. The early use of stone was entirely independent of metal tools: stone houses and walled cities existed before the Bronze Age. Masons, however primitive they may be assumed to have been, were among the earliest artisans.

Dolmens

Primitiveness is by no means synonymous with crudeness, and some of the ancient stone structures are very pleasing to the eye. The observation that nature, the textbook of our forefathers, forms smooth harmonic lines, applies very well to these constructions. Stonehenge, the various dolmens, inukshooks, menhirs, the Easter Island sculptures and many, many others all bear witness to this observation. And man-made monuments in no way outshine natural sculptures such as the Giant's Causeway in Ireland.

First quarries

Moai head made of tuff: Easter Island

The use of natural stone, boulders and rubble did not require any quarrying: the material was simply picked up. Quarrying, as we know it now, only started when man began to use metal tools and, in general, very little has changed. Fire-setting is referred to in the book of Job and was observed by Stukely in the village of Avebury in the eighteenth century for roughly shaping stone by lighting fires along the line of the intended break, and pouring over cold water while beating it with heavy stones. This method is still practised in Asia and Africa, in particular to break up stubborn rock masses, but it has been largely replaced by more modern methods. Hammers, chisels and various shaped levers, whose efficiency has been proved over the centuries, are still used side by side with power tools. The wooden hoists of the ancient builders have been superseded by metal rigs, using the same principles in a more efficient manner.

Few ancient quarries invade the landscape in the classical countries in contrast to many recent quarries, because the ancients took far more care to blend the quarries with the land. Quarry benches were hewn in to follow the natural line of the terracing.

Testaments to stone Many ancient structures all over the world testify to the durability of stone as a building material. Of the seven wonders of the world, only one, the Hanging Gardens of Babylon, is *not* made of stone, and outstanding buildings made from other materials are rare. Much of our knowledge of the oldest civilizations is based on their use of stone, without which their heritage may have sunk into oblivion. Stone is responsible for preserving the Egyptian mummies and has perpetuated the splendour that was Rome. Without marble, ancient Greece would not have been preserved for posterity. Much of modern Rome, Athens and Jerusalem is built from the very same stone used by ancient Romans, Greeks and Hebrews.

The Herodian masonry in the Temple Compound in Jerusalem has closely fitting joints, even when some of the blocks weigh eighty tons, to give the appearance of one mass of stone. By the time that Solomon built the Temple, the art of masonry was already an accomplished industry. The Book of Chronicles mentions eighty thousand quarrymen employed by King Solomon; it is likely that they used premetal-age methods as there was 'neither hammer nor axe, nor any iron tool heard in the house'. The legend that Solomon used a worm, the Shamir, to cleave hard stones 'instantly and noiselessly', might have its origin in the existence of rock-boring snails.

In Peru, the Incas built walls without mortar, achieving extremely tight joints with the help of

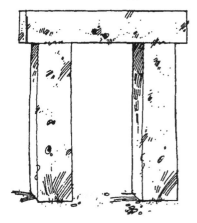

Post and lintel

lichen, not worms! It has been postulated that stones were cut to obtain the closest possible fit, then covered with plaster prepared from a lichen which still grows on the stone. After the stones were laid, the lichen eroded the remaining irregularities where oxygen was to be found, until the joints between the stones became so tight that no air could reach the lichen, which died, giving a perfect fit.

Masonry dressing has even had military significance. Bosses were left, at best to weaken the force of the ram, and at worst to give the effect of military strength with a forbiddingly rough finish, so as 'to discourage the intending aggressor', as Aristotle put it.

Many civilizations have contributed to stone architecture. The Cyclopean fortress wall at Mycenae is a perfect example of the use of undressed stones. The sculptors of Paros seem to have been the first, in 3000 BC, to have worked in marble as we know it today, preceding Carrara, present capital of marble-working. The Egyptians contributed the post and lintel structure. The Aegeans of Crete gave us the vault and corbelled arch, not knowing the true arch. The Greeks, masters of proportion, provided the columns and lintels. It was up to the Romans, the superb engineers, to construct domes and develop the true arch, first found in Persia.

In south-east Asia there is the Borobudur Buddhist Sanctuary (circa ninth century) north-west of Jogyakarta, Indonesia, which is built of andesite. There are well-known examples from India: the Konarak Sun Temple (thirteenth century), built in the so-called khondalite stone, marks the culmination of the Kalinga style in architecture and sculpture, with its exuberance of carving. And, of course, there is the Taj Mahal. In Mexico, there is the Bonampak limestone temple, an important Mayan religious centre, dating from the seventh century AD.

The post-classical and medieval times saw a major decline in stone use; ancient ruins became stone sources for religious buildings. Decorative mosaics were a major use of stone until the

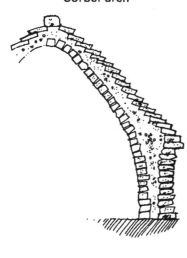

Corbel arch

Renaissance, when marble heralded the dawn of the 'New Stone Age'. Use of stone during the Baroque, Habsburg and Napoleonic eras all led up to the industrialization of stone working.

Advantages of stone   In many countries, building stone was the main construction material for urban dwellers until the beginning of this century. With the advent of modern mass-building it had to give way to concrete. The easily accessible materials of traditional building were replaced by a variety of products more suitable to controllable industrial manufacture. Dimension stone was an early victim of the use of modern, processed construction materials. Some developments, especially so-called 'cast-stone' or 'reconstructed materials' are almost indistinguishable from natural stone when first used, but after a few years of weathering they are generally better described as concrete.

Yet quite a few cities have been able to maintain their stone character to varying degrees. Granite is widely used in Oporto. Jerusalem still has a by-law requiring stone to be used for façades. Kingston in Ontario, Canada, continues to deserve its description as 'The Limestone City', and in the 'Granite City' of Aberdeen, Scotland, reconstructed stone using crushed granite has joined the solid granite frontages. Although Zimbabwe may not be known for its recent use of stone, it means 'houses of stone'; the material was used extensively in the past, by the Shona, for example. The stone town of Zanzibar is a major urban complex constructed in coral stone with important government buildings built of sandstone. Sydney in Australia has been built from sandstone which underlies almost the whole of the Sydney district.

The stone resources available world-wide meet many of our construction requirements. However, conversion into cement and other products is necessary where specifications and timetables require. But artificial, and often arbitrary, substitutes such as metal, plastic or glass have become costly to produce. Several factors are contributing to the rise of a modern stone age, the most prominent being energy saving, environmental considerations and technological developments.

Stone, in addition to its use as dimension stone, is a basic raw material and primary commodity. In fact, stone quarrying is the largest extractive industry by volume, with coal and petroleum trailing behind, and iron ore, salt and sulphur very much lower on the scale. Varieties of limestone and dolomite form the bulk of stone materials in use, and are available at or near the earth's surface over at least 10 per cent of continental areas.

Perhaps because of its widespread availability, stone has been taken for granted or even ignored. In many places, bricks, cement and other processed building materials are used where ample stone deposits exist *in situ*. A typical example is the story of a district court-house in the regional capital of a developing country which imported clinker for cement-making. The regional capital is built on an easily extractable deposit of sandstone, but instead of extracting this to provide the required dimension stone, the cement for the courthouse was transported some 480 km.

This case shows what may result from the lack of integrated planning related to stone resources. Few countries, whether developed or developing, have made an inventory of their stone resources, yet increased construction activities emphasize the need for the identification, inventory and classification of stone deposits. The data gathered from these exercises would help in the intelligent allocation of the resource, preventing such occurrences as roads being built from high-value crushed marble.

Following its extraction from the solid mountain, sophisticated technologies are applied to stone to provide good-quality aggregates or clinker for cement. This is followed by conversion into the final product, often with the help of computerized programming. And all this effort

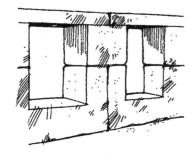

Lintel blocks in South America

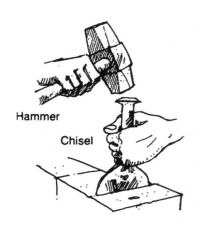

Hammer

Chisel

takes us back where we started: the production of concrete, cement blocks and pavement slabs, *to simulate the original solid stone.*

While many manufactured construction materials have gradually increased in price because of inflation, the price of stone *in situ* has been influenced by increased energy costs incurred during extraction. These aside, all other increases, including higher labour costs, have been offset by technological improvements, larger earth-moving capacities, better planning and waste disposal, and the application of appropriate technologies. For the production of cement and other materials that have to some extent replaced stone, the fuel component may range from 25 to 70 per cent of mining and transportation costs, depending on fuel inputs, transport costs, and processes used. In many countries, cement products are costlier than dimension stone, and stone units can frequently be produced at a quarter of the price of cement blocks of equivalent dimension. Cement is not the only large energy consumer; brick-making also requires sizeable calorific inputs to produce construction materials of lasting quality. Bricks are only competitive with stone where brick-making materials are available more economically than stone resources. In addition, brick-firing using wood is a major contributor to deforestation, if uncontrolled, and the construction of an average house may require many tonnes of wood to be burned.

There are other factors which favour the use of stone. Stone materials, usually obtained from the surface, require comparatively little capital investment or complex extraction techniques. The labour-intensive working requirements, and the need to use lower-grade material locally, have given stone extraction and processing a local rather than an international image. Only in recent years has a rise in international trade in crushed stone been reported. With dimension stone of a high unit value, especially marble and granite, much of the interna-

'Voussoir' arch

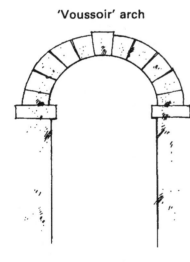

tional trade volume is now re-exported after processing.

Another advantage is that stone can be recycled and reused without causing scavenging problems. The maintenance cost of stone structures is low compared to that of maintaining structures built of most other materials, whether mineral, metal, plastic or even natural fibres such as timber and other forest products.

Experience has shown that the transition time from prototype systems to modern production methods in stone development is shorter than that of most other extraction industries. Comparatively simple extraction techniques, labour-intensive practices, and low investment requirements contribute to this. The quality of an aggregate remains similar whether the stone is broken by hand-held hammers or by powered crushers.

Intersecting vaults

The future of stone

Stone is making a huge come-back at present. The need for larger quantities of stone and the shift away from traditional sources such as granite, slate and marble, require tighter controls to be imposed on material quality and on the environmental effects of quarrying. Regular updating of standards and codes of practice is essential. Data available at the quarry must reach fabricators and finishers if the stone is to be used and laid correctly. Geotechnical knowledge has not kept pace with the new sources and varieties which are being quarried. This needs to be addressed if major failures are to be avoided.

Air pollution, a product of current times, creates a need for maintenance of stone buildings. Expert advice is essential, especially for ancient buildings, as incorrect maintenance (such as cleaning solutions) can cause irreparable damage.

With expanding use of stone, specialists in all aspects, from the location and rehabilitation of quarries to the types of stone required for modern domestic or industrial use, must have an increasing role in its future development.

This temple lion from Nepal (below) and the remains of the Citadel in Jerusalem (above), both centuries old, bear witness to the durability of stone.

One of the greatest advantages of stone is that it can be shaped using a hammer and chisel (above). Once squared, blocks can be used to build simple buildings, such as this dry-wall dwelling from Argentina (below).

2 Stone materials

Terminology

The term stone is often confused with rock, which to the geologist is any naturally formed aggregate or mass of material constituting an essential and appreciable part of the earth's crust. To the geographer and morphologist, the word rock signifies a peak, cliff or promontory, such as the Rock of Gibraltar, while to the engineer rock denotes a firm and consolidated substance that cannot normally be excavated by manual methods alone. There are other uses for the term rock, mainly of a local nature: in some mining areas, rock denotes crude ore; in quarry working, massive stone is called rock, and the list goes on.

What then is stone? Rock becomes stone after extraction, especially after diminution, that is, breaking up. The term stone is applied commercially to all natural rock materials which have been quarried or mined for constructional and sometimes industrial use. Construction stone is intrinsically a hard and consolidated material, whereas soft clays require processing to become hard, lightweight aggregates and limestone, shale and marl need to be converted to form cement and so on.

The lack of agreement on basic terminology can lead to all kinds of confusion. Industrially, the term marble is applied to a wide range of stones capable of taking a polish, although such stone can be unpolished, as in the case of travertine. Many types of travertine have properties akin to building stone and there are building stones on the market which, if polished, would satisfy the most exacting definitions applied to marble. In its wider sense, dimension stone (including marble and other rock varieties) has surfaces which are coarse and often uneven. However, definitions applied to dimension stone differ from country to country and even from region to region.

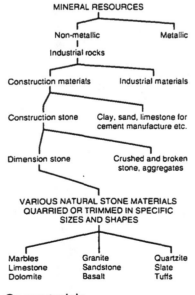

Geomaterials

[10]

For general purposes, a simplified classification of types of stone is adequate. Knowledge of the origin and composition of stones and other geological factors become important for building, and a good knowledge of stone technology or experience in the use of certain types of stone is required.

Rocks can be generically classified into three types: igneous, metamorphic and sedimentary (see below).

Some physical properties

Compact is the term applied to a rock containing slightly more than 1 per cent of pores, a figure estimated by the rate of water absorption. Compact fine-grained homogeneous rocks give the smoothest surface during dressing.

Hardness is not easy to define. It depends upon the nature of the individual grains as well as crystallization, and can be taken as the resistance to abrasion.

Fracture depends upon granularity, the size and number of pores and the natural breakage of constituent minerals, which in turn determines the degree of unevenness and roughness.

Fracture and hardness are important considerations influencing the best way of dressing a particular stone. Subsequent roughness after mounting also depends on wear: stones with well-cemented grains show little wear, whereas grains of similar hardness but with a weaker cementation suffer greater wear and thus obtain a rougher surface.

Texture is an important factor, as a coarse-grained stone or one containing shells or shell fragments prevents the cutting of sharp arrises and clean detailing.

Colour changes after dressing are, at worst, only skin-deep. Except in very dark stones, changes in hue are an advantage as they have a mellowing effect. A dressed face is always a lighter colour than the natural fracture. Some varieties of stone tend to fade with prolonged exposure.

[11]

Rock types

Igneous rock

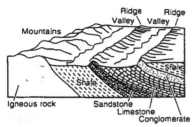

Many landscape features originate through the slow erosional removal of weaker rock, leaving the more resistant rock standing as ridges or mountains.

Most rocks, including sedimentary and metamorphic, originate from the igneous group. Igneous rocks come from molten mixtures of minerals, known as magmas, which are found deep below the earth's surface. If they cooled beneath the surface they are termed intrusive rocks, of which granite is the most familiar. Extrusive rocks, such as basalt, are those which burst through fissures in the earth's crust and cooled above ground. True basalt is usually dark and consists of calcic plagioclase and pyroxine. Igneous rock forms about 95 per cent by volume of the earth's crust (up to 16 km deep), compared with 4 per cent for metamorphic and about 1 per cent for sedimentary rock.

Igneous rocks are being increasingly used as dimension stone and are often denoted as granites or basalts, especially in the stone trade, although some of the best-known varieties are actually gabbros, norites or diorites.

Metamorphic

These are pre-existing rocks recrystallized by heat, pressure and chemical fluids. Limestone becomes highly crystalline marble; sandstone forms a highly indurated quartzite, and clays and shales become slates. Commercially, the use of the term marble implies any calcareous rock capable of taking a polish.

Sedimentary

These are ancient sediments, compacted and naturally cemented. They originate mainly from the disintegration of igneous rocks of pre-existing sediments; although some are of organic origin or formed by chemical precipitation. Sediments are predominantly deposited on the sea floor, although terrestrial sediments also occur. When 'freshly' deposited, the material is consolidated by its own weight, or the weight of other material lying on top. The cementation into consolidated material is effected by minerals carried by percolating waters.

These sedimentary rocks, including limestones, dolomites (both carbonates) and sandstones (mostly silicates) which have been

[12]

Quartz sand

1/3 each

Half sand
Half shale

Half sand
Half limestone

Shale

Half limestone
Half shale

Calcite
limestone

Mineral composition of some
sedimentary rocks

sufficiently hardened and consolidated, can be used as structural stones. Generally, consolidation is partial in a soft stone such as chalk, and more complete in a harder stone. Each bed represents a period of uninterrupted deposition. In practice, the thickness of beds ranges from a few centimetres to many metres. The carbonates form approximately 10 per cent of the exposed sedimentary rock. A similar figure has been estimated for sandstone, whereas shale forms about 80 per cent. Sandstone and shale together cover about 75 per cent of exposed sedimentary rock.

Sedimentary rocks

Sedimentary rocks used include calcareous rocks (which can be subdivided into limestones and dolomites and dolomitic limestone) and sandstones.

Limestone

Limestone used for structural stone can be divided into hard and soft varieties. The prefix crystalline is reserved for metamorphic limestones. Included in the soft limestone category are soft and hard chalks and caliche, which is not as widely used nowadays as it was in the past. Some of the hard chalks are included in this group for generic and lithological reasons, although they have the physical properties of hard limestone. Hard limestone varieties generally have a smooth, fine-grained texture with a shell-like fracture. They are usually ivory-buff to grey in colour. Deposits exploited are usually obtained from well-bedded thin layers that need no trimming on the bed, or thicker layers which may reach more than a metre. Hard-bedded limestone usually breaks well and has a rock-faced finish when cut perpendicularly to the bedding direction, especially when the direction coincides with a major joint.

Dolomite rocks

These include both soft and hard varieties: the latter are usually preferred. In most cases, dolomites are the product of a higher degree of crystallization. They are harder than limestone and may be more porous at the same time. The colour usually ranges from dull buff to grey and they are more difficult to polish than limestone.

[13]

Sandstones

Sandstones are actually hard sands containing varying amounts of siliceous (quartz) materials and cementation such as quartz sandstone. They have been classified into a number of subvarieties including graywackes, arkoses and ganister. Calcareous sandstones have been used extensively in building and construction due to their good workability and, to some measure, to their widespread distribution in more recent geological formations, often conveniently situated along coastlines. This type of sandstone generally consists of sand or shell fragments, around which calcium carbonate is deposited, at times producing an oolitic effect. Hardness depends on the amount and the nature of the calcium carbonate. Many buildings in coastal areas all over the world prove the good weathering resistance of the calcareous sandstone to salt corrosion. However, its susceptibility to sulphur dioxide in the atmosphere has been reported when used inland; for example, Cologne Cathedral.

The quality of stone materials

Although the characteristics of stonework depend much on dressing, cladding and other fixing techniques, it is the quality of the stone material that is paramount.

Seasoning

Seasoning involves the drying out of moisture. Opinions are divided as to the need for seasoning stone before it settles. It is generally agreed that stone is most easily worked immediately after quarrying. Seasoning applies mainly to soft limestone. Hard limestone seems to be less affected by quarry sap or quarry water – a condition which occurs when the sap carrying mineral cement is drawn by capillary action to the surface of the stone and evaporates, depositing a skin. Dehydration during seasoning is irreversible and subsequent artificial saturation of building stone only lowers its compressive strength. The increase in strength brought about by seasoning building stone is explained by the redeposition of percolating minerals which form a 'skin', and the simple removal of water.

There seems to be no proof, by observation

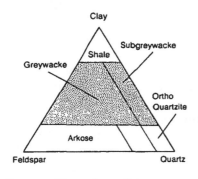

Composition of sandstones

or otherwise, that dressing either wet or dry stone after the quarry sap has evaporated, has any effect on its 'durability'. Similarly, re-dressed, previously used, weathered stone does not deteriorate any more rapidly than stone that has been recently quarried, whether seasoned or unseasoned.

Weathering

The term weathering does not essentially denote undesirable changes to stone; it may even enhance certain qualities. Decay is used to describe the ugly effects, and in most building stones it has three main causes: the effect of acid in air and rain-water, caused mainly by sulphur dioxide derived from the combustion of the sulphurous constituents of fuel; the effects of warming and cooling water in those parts of the structure that are subject to water expansion under certain conditions, usually horizontally exposed elements; and finally, the effects of the crystallization of soluble salts, introduced into the stone from extraneous sources, such as mortars.

Of the major natural causes of erosion known in physical geology, water is the main agent in the decay of stone. Rainfall limited to a few months in the year may nevertheless be heavy. Rain falling throughout the year can be damaging by its continuous action. Lichens play a much less devastating role, and damage caused by boring animals occurs only under special conditions, such as on piers and underwater structures.

Weathering problems are more prevalent in countries with extreme climates than in countries with moderate-to-warm climates. Even in the latter, however, it can be very cold in mountainous regions, causing weathering effects due to frost.

When one is considering ways in which to minimize the effects of weathering or decay, the structure of the stone is more important than its chemical or mineral composition. Uniform stones without blemishes should be selected, with no soft intercalations or seams. Soft travertines in ancient buildings have withstood the test

Granular disintegration

Exfoliation

Block separation

Shattering

Rectangular blocks
converted into rounded form
by weathering

of time, whereas many a hard granite has experienced flaking. Stone should be selected with regard to meteorological variations: some rocks may decay when exposed to the sun, yet remain in perfect condition if placed in the shade. It should be noted, however, that occasionally shade may encourage microbiological action, or the retention of damaging moisture. Features subject to weathering include vents and fissures; however, these are usually too thin to cause any strong marking. Fissures known as shakes or 'glass seams' frequently stand out on weathered surfaces.

To measure a material's behaviour with respect to weathering, stone samples are tested for compressive strength, specific gravity and porosity. Other similar samples are subjected to simulated attacks by frost, acid and atmospheric pollution; these are then given the same set of tests and the change in values gives a measure of the material's environmental stability. To date, no tests have been standardized which define a material's 'durability' when subjected to wear; for example, stone steps.

Insulation

The insulation properties of stone are less favourable than those of many other building materials. The performance of the material depends on its use and how it is affected by environmental factors. Coolness in summer is caused by the great mass present in a solid stone wall which has a high heat capacity and is therefore slow to heat up or cool down. Stone's sound transmission value is low; it is a hard and heavy material with good sound insulation properties.

Porosity

There is a tendency to overemphasize the importance of porosity in building stone. It seems to be certain that the total porosity bears no direct relation to the weathering resistance of some materials, such as limestone. Many constructions in the Middle East have been built with porous stone, as has the Coliseum in Rome, which is built of travertine, and all have withstood the rigorous tests of time. The oolitic

Ketton Stone in Great Britain is very porous compared to other oolitics, yet has proved itself to be most durable. The one-time reluctance to use porous stones for fear of dampess has also lessened with the use of cavity walls and damp-proof courses, and more care with pointing and mortar applications.

What is important is the shape, size and nature of pores, especially the degree of microporosity, defined as pores of diameters less than 0.05 mm. However, individual pores may be tubular like a capillary tube, nodular and feathering out between nodules, or thin, intercrystallizing tabulars opening 50 to 100 times as wide as they are thick. The action of carbonic acid, an important solvent for limestone, increases with greater microporosity; capillarity prolongs the dissolving action and the water is retained longer in small cavities. Thus, a large amount of micro-pores can lead to an appreciable increase in the expansion of a material. Sandstones containing colloidal minerals as a cementing medium have the most pronounced expansion. Although the more violent expansions are responsible for decay, expansion movement in pores may cause cracking, especially noticeable in smooth surfaces. The extent of frost action caused by porosity has been questioned. Sandstones with a higher expansion coefficient than limestone are most resistant to frost and it is unlikely that unequal mineral expansion alone can account for rock burst.

The Coliseum, Rome

Florescence

The term florescence is used to describe both efflorescence, which is the appearance of salts on the stone's surface, and cryptoflorescence, denoting crystallization within the pores. Colloquially, especially in the building trade, efflorescence is also known as wall white, stack white or wall cancer. The florescence phenomenon is due more to the careless use of setting materials than to the stones used.

Salts crystallize externally if evaporation takes place from the surface. If evaporation takes place below the surface, then the salt is deposited internally and changes in molecular volume take place which may result in decay, which is more harmful than efflorescence.

Little quantitative data is available on the extent of the damage caused by sea salts, which remain deep below the surface and are not easily washed out.

Strength

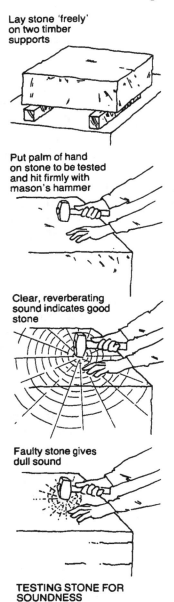

Lay stone 'freely' on two timber supports

Put palm of hand on stone to be tested and hit firmly with mason's hammer

Clear, reverberating sound indicates good stone

Faulty stone gives dull sound

TESTING STONE FOR SOUNDNESS

Basically, building stone can be affected by three types of stress: compressive stresses which tend to decrease the volume of the material, causing shattering; shear stresses, which move one part of a stone with respect to another, under certain conditions inducing a permanent change of shape; and lastly, tensile stresses, which produce cracks and fissures; torsion (or twisting) is less common.

Since, in the majority of cases, stone is now being used for facing rather than for foundations, piers or bearing blocks, its strength should be tested only in specific cases. At the same time, it has to be borne in mind that laminated stone, suitable for solid masonry, may undergo tensional stresses when used for cladding.

The mechanical properties of rocks depend not just on the sum of the strength of constituent minerals, but also on cementation and other factors, and cannot be determined simply by adding up the properties of constituent minerals. The bond strength of mineral grains and the bonding cement are important factors.

Not all methods for testing strength are universally accepted and standards vary. Results obtained from tests on stone from the same quarry may show a variation of 50 per cent. With such a deviation, a simple test on the stone from any quarry would seem to have little significance. (See Appendix: Stone testing.)

Strength tests, whether for resistance to crushing or resistance to direct strain (tensile strength) do not take into account internal stresses resulting from changes in temperature, frost and other variables. The behaviour of the stone in several testing sitations may yield evidence more valuable than the numerical results of mechanical and other tests.

Advanced technical testing procedures are beyond the scope of this book, but proceedings are outlined briefly in the Appendix. It should be noted that a standard sample of about 10 cm^3 of rock is not always representative of a rock deposit which is usually more than a million times larger in volume. Modern subsurface logging techniques (sonic, density, neutron), are made *in situ* and give more representative results.

The highest crushing or compressive strength is obtained when the stress is normal to the bedding. Saturation decreases the compressive strength of the rock, as the incompressible pore liquid under stress may inhibit the normal process of movements. A strength reduction of 10 to 12 per cent in limestone and 50 per cent in sandstones has been observed.

Transverse strength is very important. Great strains may be induced by shearing or bending stresses, caused by foundations settling unevenly.

Tensile tests have given higher results for limestone than for any other group of rock apart from slate, but still give a low figure when compared with concrete slabs, for example. Generally, strength figures have been obtained by taking average values for a large number of well-proven building stones. The classification of crushing strengths is useful for comparing different stones. In a similar manner, the strength ratios for particular stones can be obtained.

Crushing strength : tensile strength = 10–15:1
Crushing strength : shearing strength = 20–40:1

Fine-grained rocks have a tendency to be stronger than coarse-grained. Rocks with interlocking between the crystals are stronger than rocks with poor interlocking. The elastic properties of stone may be important, for instance in earthquake-prone areas. A fine-grained limestone may reach a constant of elasticity amounting to just below 10^6 kg/cm^2.

Weight Sometimes the weight of stone is important. For the building of a quay or dock, walls of a high density are essential as the stone is immersed in water and loses a considerable part of effective required weight. On the other hand, vaulting requires a light stone. Several hundred varieties of stone were tested for various constructional purposes and their specific gravity was found to range as follows:

Soft limestone	1.64–2.15
Hard limestone or dolomites	2.36–2.75
Sandstones	1.75–1.95

These figures are only representative of the group investigated.

Hardness Hardness is not an essential quality for stone used in facing. However, it is a requisite where the stone is subjected to abrasion as in steps, door sills, paving and flooring. A hard stone gives forth a clear metallic sound when struck with a hammer.

Specifications for building stone

The general requirements for stones used in building can be summarized as follows:

o Sound, uniform rock material.
o Presence of rift to facilitate workability by hand tools.
o Porosity – advantageous, providing it does not influence resistance to weathering.
o Chemical stability to prevent florescence.
o High strength requirements – only in exceptional cases.
o Low specific gravity – desirable for easier handling, such as for roof vault lining.
o Low abrasion factor – for use in paving, flooring, sills and so on.

The use of slate need not be confined to roofs! Above: *Laying a slate damp-proof course, Nepal.*
Below: *Stone that was edge-bedded during construction has suffered from weathering, while the naturally bedded stone surrounding it is in far better condition.*

Hardness is an important quality in stone which is subject to abrasion.
Above: *Stone for paving the streets is transported from Kholegaon quarry into Surkhet Town, Western Nepal.*
Below: *Stone staircase on the Nepal-China Kodari Highway.*

3 Stone sources

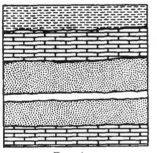 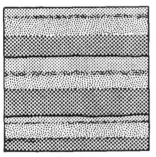 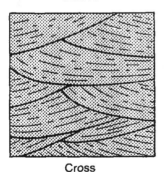

Locating stone

FORMS OF BEDDING

Regular

Graded

Cross

Stone was in widespread use long before geology became a formal discipline. If a formation seemed to be sound and 'block-like' in appearance, then stones were selected from rock outcrops and shaped for various uses. Boulders transported by glaciers were used, too, together with other loose rocks. Slabs and blocks of various stone types, including marble, limestone, sandstone and the uppermost sheets of granite outcrops were worked to easily accessible depths. Eventually, these turned into quarries, until national planning controls restricted their development.

Few countries, particularly in the developing world, have records of quarries, whether in current use or obsolete. The use of basic geological maps greatly facilitates the location of stone deposits, immediately indicating areas where particular stone types are likely to occur. Surveys of rock formations, also useful, are sometimes available at sites of geological surveys, geological departments of institutions and technical colleges and from consultants. Regional or local literature dealing with nature, topography, field lore and other similar subjects are worth consultation, too. Such works may contain valuable descriptive information on local stone resources, even if they are out of date for use as teaching texts.

Much information can be gleaned by reading the legends of maps and explanatory publications, which often indicate whether a material is soft or hard, whether the deposits are well-bedded or laminar and other pertinent points, as well as giving the lithology of the deposit. Prospecting in these areas, paying special attention to surface features, can indicate whether more detailed exploration is worthwhile.

The establishment of an inventory of existing

and potential stone resources is invaluable, not least in considering environmental factors such as urban planning, drainage problems and road construction. Stone resources provide an economic source for building material, provided that the market is situated near to the deposit. It cannot be emphasized strongly enough that building programmes, particularly low-cost housing projects, should be sited close to stone deposits to obviate the necessity for special access roads, the costs of which often exceed those of the development of the source.

Types of deposit

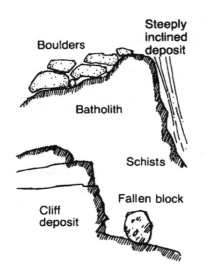

The choice of material for stoneworking depends primarily on local availability, followed by various criteria, headed by ease of extraction. Limestone is comparatively easy to work and is used widely, especially the soft varieties which form a harder skin after prolonged exposure to air. Volcanic tuffs with similar properties have proved their stability in use.

Unconsolidated boulder deposits of the fieldstone and river-stone types, glacier accumulations, mountain screes, stones from dry riverbeds, terraces and beaches have all been used since time immemorial to provide material for house building and various other constructions. These types of deposit have generally been formed by streams, and have an additional advantage that in many cases natural sorting has taken place, and only strong and stable stones have survived periods of transportation and rolling which may have taken millions of years.

Sometimes, such deposits are found as gravel beds or terraces, in which case, selective excavation will be required. Often, screening by setting up a sieve in the field is sufficient to obtain the sizes of stone required. However, special care has to be taken as softer rocks may be present, preserved when the terraces were formed. These may have been weakened by constant humidity and other weathering action, resulting in the loss of cohesive properties, or they may have been affected by active

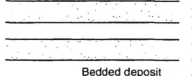
Bedded deposit

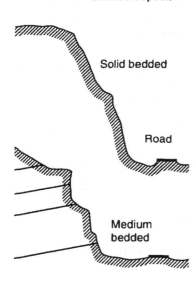
Solid bedded

Road

Medium
bedded

chemical and bacterial reactions, such as acid or humic soil altering their minerals.

All these types of deposit are categorized as boulders or 'float', even when they are not strictly of the usual size or shape. They may contain a wide range of rock-types. Loose stone may also be obtained from disused quarry and mine waste-heaps.

Stones with flattened profiles are the most practical to use. The quality of dry walls and other structures built from field-stones depends on the way in which the stones are stacked. Often, no tools are required other than those used to dig foundations or level the ground. In other cases, hammers are the only tools required for trimming or sizing stones to the required shape or size. Rubble walls and random courses can be built from boulder deposits.

Bedded deposits, usually of the sedimentary variety, are next in importance for ease of extraction, although the efficiency of extraction depends on the tools available. These deposits have many advantages over boulder deposits, as sizes and shapes can be predetermined and formed as required during extraction. Slabby deposits with layers between 6 and 8 cm are the easiest to work, as they can be extracted and lifted with various types of iron bar, then broken into shape *in situ*. Thicker layers are usually subdivided with pneumatic tools and wedges or with cutting and sawing machines. Bedded deposits of suitable thicknesses do not require horizontal cutting, as the blocks can be lifted at the bedding plane, thus saving much time and hard work.

Near-vertical deposits which are quarried can be sedimentary, igneous or metamorphic, and either bedded, fissile or schistose. Many slate quarries are near-vertical due to the jointing system peculiar to slates. Although excavations in this type of quarry follow easy lines of extraction, the advantage is offset by the need to go to greater depths and the difficult working conditions at inconvenient angles.

Cliffs are formed by harder strata and usually

have little overburden, as this has been eroded. Difficulties in extraction include the need to take special precautions, often requiring quarrying aids like safety belts and ropes, and also scaffolding, all of which add to the cost. Some cliffs consist of loose blocks separated by jointing and bedding and these can be removed fairly easily, subject to the precautions outlined above.

Massive deposit prospects are more typical of igneous rock-types. It is easier, if more expensive, to plan quarry development and the placement of equipment in massive deposits. Although they can be more uniform in colour, physical, mechanical and structural properties, they require horizontal cutting, unless sheeting planes are artificially induced.

Boulder resources

Cobble and boulder deposits are an important source of structural stone, particularly in informal construction. Boulders are considered to be the largest size of gravel, starting at 256 mm diameter which, assuming a perfect spherical shape and taking 2.70 as the specific gravity, should weigh 23 kg. The smallest cobble size starts at 64 mm and weighs 3.75 kg. Boulders, cobbles and pebbles are sometimes defined as unconsolidated sediments. A limited study has been made of the coarser gravels, especially for structural use, but there is still little agreement on the relative importance of abrasion, solution and shape-sorting of pebbles on their structural behaviour; nor is it clear how the investigative results from pebbles or coarse sands can be applied to large gravel fractions, as the size scale from a pebble to the smallest boulder exceeds 1:100.

In practice, it is easier to gather material from sources *in situ*. At source, be it road or river-bed cuttings, the gravel is usually in a stratified graded position. The stratification may be horizontal, inclined or the gravel may be unstratified.

Quartzite and compact rhyolite gravels are most resistant to abrasion during the geological transportation process, whereas granites are variable, and all strongly micaceous rocks, sand-

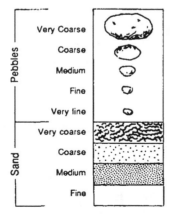

Gradation in relative size of pebbles and sand. This diagram is not drawn to scale.

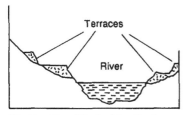

Rounded rod

Flat

Oval

Flat

Rounded blade

Sphere

River valley filled with alluvium which has been eroded into terraces containing a variety of stone materials

Terraces

River

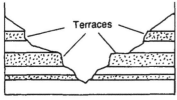

Terraces

Valley displaying more resistant strata forming cliffs and terraces

stones and limestones are easily abraded. Other varieties hardly survive glacial or fluvial transport.

The grading and sorting of cobbles and boulders is a prerequisite for structural use. Generally, sizes can be based on the largest dimensions of 10, 20 and 30 cm, and subdivided according to shape, according to whether they are long or rod-shaped, flat or blade-shaped, triangular or rectangular.

Although the general stoneworking techniques discussed in Chapters 3 and 4 apply to boulder-type deposits, it is convenient to discuss some more specific uses of this type of deposit. Where cobbles and boulders are rounded beyond easy use, they can be given flattish faces and arrises, provided that the stones are locally available and where time for labour is not the main concern. When flattish on two sides, a guillotine can be used for splitting.

The limitation of the cutting of any rounded or curved profile is the high strain sustained by any cutting tool surface, be it a knife edge or press tool, owing to the concentrated, rather than linear, pressure on the top and bottom of the workpiece. More linear contact can be achieved by cutting a straight V- or U-shaped notch or groove, using a file, carborundum edge, or cutting disc across the stone. The type of cutting aid depends very much on the properties of the stone and the quantity to be cut. Contact can also be increased by using equalizing blades.

Stone materials are subject to abrasion, grinding and additional breaking of the larger particles during transportation. The breaking tends to depend on schistosity, stratification and weakness due to differences in texture. An efficient transportation agent will tend to place together particles of roughly equal shape and dimension. Research to date has reached the following conclusions. Shape sorting occurs in the transportation of pebbles: spherical shapes tend to be rolled together; ellipsoidal shapes tend to become associated; disc-shaped particles become sorted out from others and particles with little rounding may lag behind during transportation. Abrasion is the physical agent governing the flatness of beach pebbles. Wind abrasion produces certain types of shapes. Glacial action results in flat-iron shapes, with striated surfaces. Pebble shapes of turbidites are more similar to river rather than beach deposits.

Boulders up to several metres in diameter tend to be poorly rounded, if they are rounded at all. The largest boulders in particular are transported by mountain streams, waves and the currents of stormy coasts, glaciers and mud flows on land and by icebergs in seas and lakes. Cobbles are generally better rounded than boulders, and because they have travelled longer, the survivors contain on average a greater percentage of resistant, hard and tough rocks.

Gravel deposits or loosely cemented conglomerates are governed by sorting size distribution, material stratification and grading parameters, all of which are important in determining structural applications. These factors are subject to depositional and environmental factors which can only be determined locally. Generally, gravel deposits are more sharply defined at the base than at the top. Ellipsoidal shapes tend to lie with the long axis perpendicular to the direction of flow, whereas the upper surfaces of dish shapes are inclined up-current. Gravel in beaches tends to overlap in a seaward direction.

Conventional splitting systems apply mainly to heavier boulders by drilling holes with a jackhammer and breaking using the 'plug and feather' method and controlled blasting. However, when the boulder is partly buried below ground level, only chipping can be achieved and the angle of the hole will determine the depth of the break. Rock-splitting devices such as jacks or inflatable bags assist in further opening cracks and joints already formed by breaking.

Sorting shapes In practice, each stone should have its own function, whether its largest face is exposed in cladding requiring little structural attention, or its smallest face is stacked for load-bearing purposes. Attention to primary sorting, handling and hauling is vital for the optimum availability of stones at the site, where further sorting will be carried out. Stones with right angles form the most important pile. A second pile should contain larger and flatter stones for bonding. Large thin slabs should be carefully stacked in an upright position to prevent breaking.

Identification of deposit

Unlike other mineral resources, stone deposits are relatively widespread and easy to identify when they are not covered by overburden, loose stones or heavy vegetation. Deposits can be identified by careful observation of, initially, the local use of stone materials, past and present. The first step is obviously re-examining outcrops and old workings in established stone-working districts. The nature of cliffs, banks, road-cuttings, canyons, gorges, wadis, river-beds, valleys, underground workings, tunnels and other prominent features, natural or man-made, should also be surveyed. Besides a knowledge of geology, previous experience in the development of stone resources is an advantage when looking for workable deposits. Although a more systematic stone technology has emerged since World War II, past experience is still very important.

Highly folded or faulted areas are usually avoided in the search for dimension stone deposits, as are areas where rockbursts may occur. In metallic mineralized areas, detrimental pyrites may be anticipated. Flint beds, apart from inhibiting efficient exploration, often indicate areas of intense silicification, making extraction and quality upgrading expensive. Generally, deposits requiring expensive or unconventional infra-structures should be avoided in the early stages of prospecting.

Further exploration

The extent of geological investigations depends on the requirements of the prospector. Geological mapping and sampling are fundamental to determining the quality and consistency of the stone and the quantity available. The samples, taken either by simply breaking off pieces of rock to examine the unexposed surface, or by core drilling while proving reserves, are checked for physical or ornamental properties. Whichever method is used, it is important to ensure that the sample is taken from the rock *in situ*. Large broken-off cliffs, some weighing several thousand tons, can be easily mistaken for solid outcrops. Sophisticated instruments are not always required, as surface features, observed by the trained eye, can reveal much about a stone deposit.

Geophysical investigations in stone evaluation are usually limited to the determination of overburden, the presence of a water-table, undesirable soft rock layers and compactness of beds. Core drilling, besides determining reserves, will confirm the presence and extent of some of these features; methods exist to correlate defective zones and the position of cracks, cavities, leached features and broken-up and open veins in the boreholes. Good core recovery usually indicates a satisfactory quality. In addition, a whole range of subsurface logging equipment and methods are available, including sonic, density and neutron techniques, which are applied in industrial investigation.

Variations in the deposit over relatively short distances may inhibit its economic viability, not only because of planning or extraction considerations, but also because continuity in the supply of identical material for extension purposes and repairs is of great importance when a planner, architect or builder chooses his or her material.

There is no single answer to whether the physical condition of the outcrop reflects the quality of stone underneath. The depth of weathering, for example, can vary greatly with climatic conditions and the nature of the rock, to produce a wide range of effects. Again, there are outcrops where only the stone seen on the surface is of economic value, as in the Arad area near the Dead Sea. Defectiveness may increase with depth in deposits which are not homogeneous

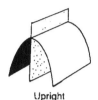

Upright

Inclined

Overturned

Recumbent

or in strongly karstic areas where joints widen below the surface; yet, in other cases, defectiveness *decreases* with depth. Each prospect has to be judged individually.

Generally speaking, outcrops of caliche or leaching features suggest an area is unfavourable for extraction. Caliche is often due to the excessive porosity of the underlying rock, and the same applies to a leached-looking surface, pointing to post-depositional chemical action. In the above-mentioned Dead Sea area, for example, the outcropping black and brown marbles are sound, but below the surface leaching or hydro-thermal action has left its mark.

At the prospecting stage, the following points are relevant:

Location. Accessibility, legal aspects and planning status of the land.

Traverse. A field traverse of the deposit should be made, to study surface features of the terrain and measure the thickness of beds, where possible.

Sampling. If the terrain is hilly, then samples are taken from each bed along a straight vertical line across the layers. Otherwise, sampling may proceed along horizontal lines and perhaps by trenching. The sample should be of a size that can be removed conveniently with a hammer and chisel.

Small-scale geological mapping. Notes and sketches should be made, on a topographical base if available, of all relevant features observed. Topography can be sketched in with the aid of a compass if a suitable base map is not available.

Evaluation of samples. This includes the examination of ornamental qualities, and physical tests where the size and quantity of the samples allow. Petrological examination is also important.

Preliminary economic evaluation. This covers potential reserves, infrastructure, working facilities, workability of the stone. The existence of a viable market for the stone should be confirmed.

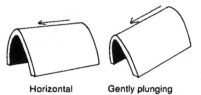

Horizontal   Gently plunging

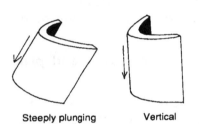

Steeply plunging   Vertical

If the preliminary economic evaluation is favourable, more detailed exploration will usually include:

o Topographical and geological mapping at 1:2500 or larger scale.
o Core drilling, to ascertain the quantity and quality of the reserves – NX size (6 cm or $2\frac{1}{8}$ inches in diameter) on a 50-metre interval grid.
o Geological, geophysical and technological evaluation of the nature and extent of the deposit, to include the distribution of the formation, its stratigraphic position, structural elements, lithological contacts, fracture systems, cleavage, overburden, weathering, colour consistency and texture and physical properties.
o Extraction of production-sized trial blocks for commercial testing.
o Economic evaluation and a feasibility study taking in the scale of the working, the availability of skilled and unskilled labour, transport facilities (roads, railways, waterways), necessary services (power, water and so on), development factors such as urban planning, market surveys and capital investment.

Surface features of stone deposits

There are basically two types of stone deposits: unconsolidated, which refers mainly to boulder deposits, and consolidated deposits, which refer to sedimentary deposits; igneous rocks such as granites and basalts often have jointing but no bedding (see below).

Bedding

Beds, or strata, the elementary units in consolidated stone deposits, are not to be confused with benches. Beds denote natural depositional entities, whereas benches are man-made to match extraction requirements. In sedimentary deposits, four types of bedding (and their variations) can be distinguished: massive or thick bedding, thin bedding, irregular bedding and pseudo-bedding. Pseudo-bedding, which is imparted, for instance, by thin clay-layered coatings, may extend into a deposit for only a short distance,

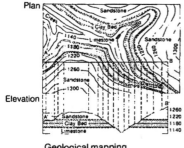

Plan

Elevation

Geological mapping

Bedding plane
Dip 50°

Strike

Sedimentary deposit

making it appear thinly bedded. Distinguishing pseudo-bedding from true bedding planes is essential, especially when large blocks are required, for which very thinly bedded deposits should be avoided. Bedding planes in massive deposits are used to advantage to avoid costly and tedious undercutting during extraction and for the establishment of convenient natural bench or quarry floors. Thin beds are favoured for building stone, tombstones and pavement slabs, after trimming into appropriate shapes.

The thickness of the various types of bed described above is denoted arbitrarily. A more formal scale of stratification thickness is shown below. All the bedding types may include conglomeratic rocks which could become unconsolidated deposits when they have a weakly cemented matrix.

Very thickly bedded	Thicker than 1 m
Thickly bedded	30–100 cm
Medium bedded	10–30 cm
Thinly bedded	3–10 cm
Very thinly bedded	1–3 cm
Thickly laminated	0.3–1 cm
Thinly laminated	Thinner than 0.3 cm

Dips A sudden increase in dip indicates a displacement, however small. In certain instances shallow dips, which are common, can be used to move the material down-dip by taking advantage of the natural gravity of the inclination, and to facilitate drainage during rainy seasons.

Deviations from regional dips indicate possible irregularities. Sudden changes may point to a fault zone, where strata are more fractured than elsewhere. Irregular dips may indicate small folds, landslides, underlying unreliable strata and many other features which render a deposit unsuitable. Highly tilted beds, typical of rough topographic features, are usually crumpled and fractured, and may not provide sound stone blocks. The dip direction affects the haulage grade, an important factor in the development of a deposit.

Joints Closely spaced joints can prevent the extraction of sizeable stone blocks, and irregular joints can cause projecting bellies which both reduce the size of the blocks extracted and add to trimming costs. On the other hand, joints usually facilitate extraction as they provide headers (natural breaks), allowing sections of the deposit to be opened.

With igneous rock, jointing may separate the rock mass into the well-known sheets seen in many granite quarries. The suitable spacing of sheets obviates the need for undercutting during extraction. The identification of this jointing is more critical in igneous than in sedimentary deposits.

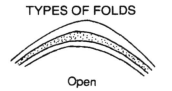

TYPES OF FOLDS

Open

Faults Loose rock or soil cover may obscure the true location of a fault. Faults can, however, be inferred from a sudden change in the topography or by changes in dip; steepening usually indicates displacement. Generally, deposits in displacement zones are to be avoided, as rocks there are usually softer, fairly porous, iron-stained and they contain a large proportion of worm holes and seams as a result of brecciation or shattering and penetrative weathering. Faults need not always be detrimental. The much-valued Peqiin marble of Galilee in Israel, a fault breccia, forms a belt over one kilometre long, some hundred metres wide, and reaches an unknown depth. Quarrymen are usually more concerned with small faults, and these are better termed 'displacements'.

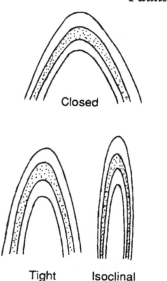

Closed

Tight   Isoclinal

Folds Horizontally bedded stone deposits are not common, although seen in quarries with greater frequency than average due to selection. Near-horizontal beds are often found on the crest of low-dipping folds. These so-called flat deposits are to be avoided, as the bending of a fold affects the crest of the anticline and the rocks are 'weaker' there than on the flanks. Small folds, often obliterated by topographical features, are usually revealed by closer examination of dips.

Rounded

Veins Veins, dykes and sills may criss-cross a deposit at various spacings and disrupt methodical extraction, if their proportions are significantly different from those of the surrounding material. If veins and the surrounding matrix differ in hardness, uneven surface dressing or polishing will result. Differential weathering can be expected when stone from such a deposit is exposed on the exterior of a building, although at times small amounts of veining may provide an interesting feature.

Angular

Karst Karstic features are present in many limestone areas. Underground channels may reach great depths, widen to form underground caves and cross a number of layers without any apparent sign at the surface. The filling of channels and cavities with soil affects the quality of an otherwise good stone deposit mainly by discoloration. Defects due to karstic features appear in some high-quality beds, and are difficult to avoid. They interfere with any regular extraction system based on the jointing pattern.

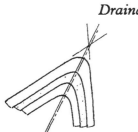
Symmetric

Drainage Work in many of the smaller deposits is affected by heavy rain. Drainage facilities are necessary to prevent flooding and to minimize waste, especially where washed-in soil contaminates the more porous strata and discolours the stone. Extraction should preferably take place on a level above the drainage channels, such as on the side of a hill. High initial development expenses are offset by lower operation costs in the long run.

Asymmetric

The priority classification shown overleaf depends on various topographical and structural factors which are schematized to give a general

[35]

idea for simplicity. Other cases, such as the absence of near-vertical jointing or pronounced folding and faulting, require examination of the prospect to determine the most economic extraction technology.

Deposit types for economic stone production*

Slabby deposits
○ Gentle dip and near-vertical jointing.
○ A large proportion of layers of usable stone (6 to 30 cm).
○ A widely spaced jointing system and little overburden.
○ As above with less favourable jointing.
○ Up to 3 metres of overburden.
○ Thicker bedding up to 80 cm thick.

Deposits without apparent bedding
○ Gentle dip and near-vertical jointing.
○ Widely spaced jointing and little overburden.
○ With less favourable jointing.
○ Widely spaced jointing and 2 to 3 metres overburden.
○ Narrowly spaced jointing and 2 to 3 metres overburden.

Slabby deposits
○ Pronounced dip and near-vertical jointing.
○ Without apparent bedding, as above.

*Ranked in order of potential

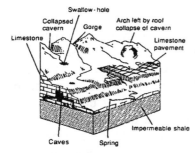

Some features of a karst landscape produced by solution of limestone

Above: *Folding of bedding at the Kotah slab quarries, India.*

Below: *Dry wall made from granite boulders, cobbles, smaller pebbles and stones.*

Above: *The well-spaced jointing of this granite quarry facilitates the removal of industrial-sized blocks.*

Below: *Another example of widely spaced jointing, this time in a hillside deposit of well-bedded marble.*

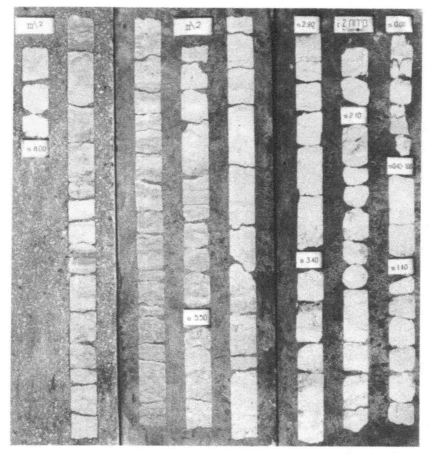

Above: *Cylindrical cores obtained during exploratory drilling of marble deposit (see page 46).*

Below: *Cuboidal jointing facilitates the extraction of building stone from this limestone quarry.*

This potential quarry site for grey marble in Haiti shows rill weathering typical of karstic action.

[40]

4   Stone extraction

Quarry planning and development

The term quarry usually implies the extraction of stone and other construction materials in surface workings. The planning of a quarry, taking into account geology, drilling results and other factors, has as an important objective the minimum loss of material. The development of a quarry is often the most expensive stage and proper use should be made of available technology. The objectives of quarry planning and development include:

o Facilitating an increase in production at short notice – due to a rush order for a large building, for instance.
o Uniformity and controlled quality of the product.
o Selective extraction of several qualities and colours at the same time.
o Conservation of rarer varieties within a deposit for special uses.
o Minimum capital tied up in stockpiling.
o Adjustment of production to reflect fluctuating price levels.
o Removal of wastes and rehabilitation of land (see Chapter 5).

Stone extraction operations can be roughly divided into three categories: pit quarries, hillside quarries and underground workings. A further subdivision may include trenches, adits and tunnels. A hillside quarry, the most common type, may develop into a pit quarry, and eventually be continued underground. High land cost, legislative factors, town and country planning problems and environmental considerations are important reasons for going underground. Underground workings which are developed year-round tend to produce material of higher quality than that at the surface, as they involve no overburden removal, greater cost

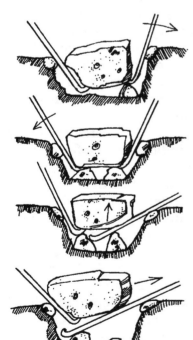

Four stages of field-stone extraction

consciousness and an extracted material less affected by weathering. However, for economic reasons, stone is extracted underground only when open workings are not feasible, as lighting, ventilation, dust collection and other maintenance factors add considerably to running costs. Pit quarries, unlike hillside quarries, require special drainage measures.

Location In the past, quarries evolved from small workings which had started out to provide stone for local use. Systematic searches began in the twentieth century, with scientific methods and appropriate technologies only being fully applied during the last few decades. An adequate knowledge of geology and morphology is an obvious requirement in quarry siting, as are economic considerations including logistics, the distance from a consumer centre, availability of labour, roads, water and power. The siting of the processing mill is important, too, and its distance from the quarry will depend, among other things, on the quality of the raw material. Where primary cutting results in much wastage, trimming facilities at the quarry may save considerable transport costs. Stockpiling is less expensive near the quarry, in that capital is not tied up in transport or storage space.

The world-wide tendency for beginners to quarry easily extractable boulders seems to be irresistible. This material is of a variable quality and its use should be judiciously confined; poor quality marble or stone found in building more often than not comes from boulders. Boulders should be carefully selected before marketing in slab form for ornamental purposes, as variations in colour, pattern and texture are common, and, with few exceptions, boulders are more affected by weathering than *in situ* stone. Inferior stone creates lasting disillusionment among architects and builders, and slows the development of local stone industries in many parts of the world.

Stone extraction legislation should have a strong environmental slant. Even ample resources require conservation and quarry leases should be issued judiciously. The control

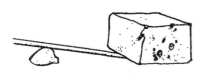
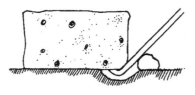

Moving blocks with levers

[42]

of explosives is essential as their use, even over a small area, can damage a sizeable deposit. Royalties or other levies should be put at a rate which does not discourage development. The market price of marble or stone is reasonably steady compared to metals, lumber or concrete.

Opening up quarries

At one time, opening up stone or marble quarries presented few problems: country planning restrictions were minimal, and sites could be chosen at random. If the quality of the stone was unsatisfactory, it was easy to move on to the next convenient site with no worries about reclamation. Available knowledge was not always sufficient to develop deep quarries and most workings were shallow; hence the numerous holes which can be seen dotting any stoneworking area.

Increased environmental awareness, the necessity for larger workings and a general need for efficiency, with special emphasis on finding enough proven reserves to justify considerable capital outlays, have changed the picture. Restrictions on opening new surface quarries might force operations to be started underground, particularly in densely populated areas where planning authorities may have earmarked all available land for other purposes, ignoring the need for quarries.

The tendency to open conveniently accessible quarries can result in bypassing stone deposits with considerable potential and concentrating instead on those of lesser potential. Such an approach, especially in undercapitalized countries, has often had unfortunate consequences for over-enthusiastic entrepreneurs exploiting the first available prospects. A nationally integrated and controlled approach, giving high priority to making an inventory of available resources, will reduce mistakes of this kind and ensure more rational use.

National planning and inventory

A stone resource survey, used as a base for purposeful planning, should cover conveniently selected regional units. The inventory should

STONE-BREAKING

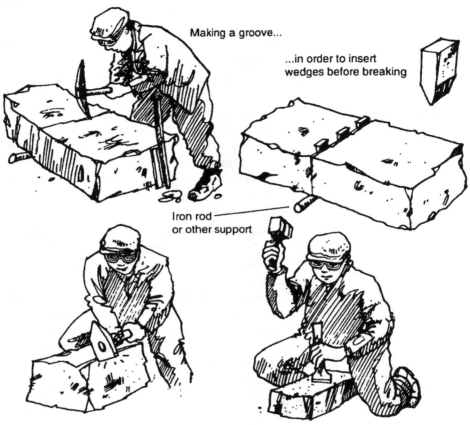

Making a groove...

...in order to insert wedges before breaking

Iron rod or other support

Breaking an edge with a hammer... then trimming with a chisel

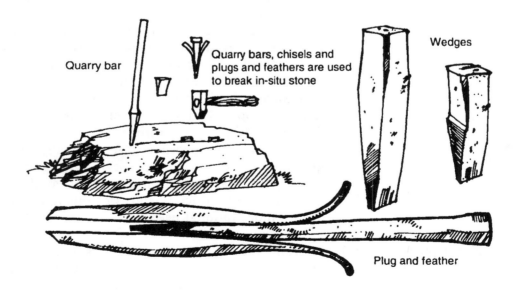

Quarry bar

Quarry bars, chisels and plugs and feathers are used to break in-situ stone

Wedges

Plug and feather

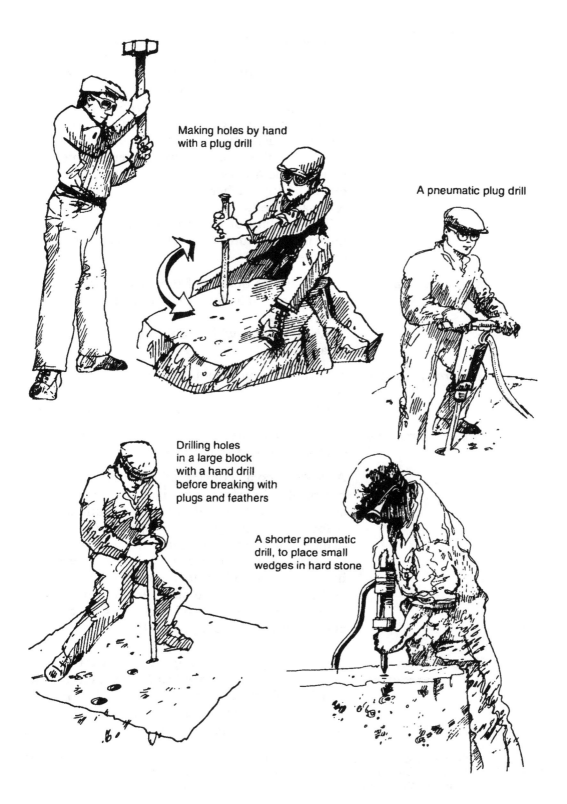

Making holes by hand
with a plug drill

A pneumatic plug drill

Drilling holes
in a large block
with a hand drill
before breaking with
plugs and feathers

A shorter pneumatic
drill, to place small
wedges in hard stone

[45]

include not only descriptive and factual data on the resources, located and illustrated on maps of a suitable scale,[1] but also notes on relevant planning requirements and development needs.

The starting point is the collection and review of all existing records within the archives of national institutions responsible for geology and mineral resource development. This data will provide a basis for assessing the nature and extent of any further surveys and allow priorities to be assigned.

The grading of potential resource areas must take into account a number of environmental considerations:

Aesthetic considerations
- Visibility of quarries from main roads.
- Scars on hillsides created during and after quarrying.
- Waste heaps which do not blend with the landscape.

Disturbance factors
- Dust and smoke pollution of atmosphere.
- Noise.
- Effect of excessive quarrying dust on vegetation

Antiquities protection
- Presence of archaeological excavations.
- Historical and religious sites.

Although cost considerations may limit the extent of subsurface investigations, an adequate advance knowledge of reserves is fundamental before a quarry goes into operation. In core drilling for proving reserves and assessing quality, maximum recovery of the core is critically important. A collection of polished split cores embedded in concrete slabs enables easy reference to the core log and facilitates description and classification (see page 39).

Preparation and extraction
The first step in preparing a new area for quarrying is to clear the overburden, which usually consists of unusable rock and soil. An unhampered working area minimizes the risks of falling rock or debris and, especially in pit quarries,

[1] The scale of maps used varies. A scale of 1:100 000 is adequate for general location; 1:20 000 for the first stage of the investigation, and as large as 1:1000 for detailed work depending on the uniformity and size of the deposit.

[46]

exposes the pay-rock, that is, the saleable rock, for quality assessment. The equipment to be used depends on the nature of the overburden – soft rock, hard rock, soil, overgrowth or alluvium – and also on the topography; the choice may include bulldozers, shoveldozers, rippers, high-pressure water hoses and power shovels. If labour-intensive methods are to be used, then overburden can be removed with hand-tools. Explosives are used only if it is certain that the deposit will not be damaged.

The development of a quarry depends on the nature of the deposit. Modern practices require shallow benches, with the thickness of deposits, vertical changes in beds, and economic considerations determining their height. The shallower and wider the bench, the safer it is. The equipment used may influence the bench size: the use of traditional wire-saws favours the maximum height, whereas with jack-hammers the height is limited to ensure straight cuts. Waggon drills, quarry bars, chain-saws, modern diamond wire-saws and their variations are used for medium-height benches. The choice of extraction equipment also depends on the nature of the deposit.

Quarries for building stone should be in well-bedded deposits where layers are approximately the thickness of the stone required, and where slabs can be easily subdivided. Stones from massive beds are usually produced from the softer limestone varieties and dolomites. Few sizeable quarries are known which have been worked over a number of years with the sole purpose of producing building stone unless extraction is mechanized. Usually, for building stone, shallow pits are opened and exploitation is restricted to the upper layers.

The choice of extraction method depends greatly on the nature of the deposit and the material, as discussed in Chapter 3. The same factors determine subdivision methods and the further working of the stone. Both the deposit and stone have so many variables that no simple working methods can be predetermined without knowledge of the characteristics of both. Some

BEDDING DIRECTIONS

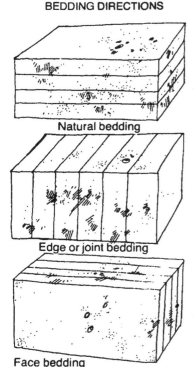

Natural bedding

Edge or joint bedding

Face bedding

of these factors are listed below. (The examples in brackets denote characteristic materials extracted from the deposits.)

Factors influencing extraction and working methods

Deposit	Stone
Unconsolidated	
o Boulder deposit (granite etc.)	o Physical properties
	o Chemical properties
Consolidated	o Tectonic factors
o Massive deposit (gabbro, dolomite, syenite)	o Colour
o Near-vertical deposit (schists, slates and quartzite)	o Mechanical properties
o Bedded deposit (limestone, schist)	
o Cliff (sandstone, limestone)	

The objective is to extract stone blocks without destroying the pay material, simultaneously developing the quarry. Extraction may be carried out in a number of ways and these are listed in the table below.

Mechanization, unless well planned, may reduce the benefits of good grain direction during quarrying, resulting in chipped-off corners during transport, poorer workability in slabbing operations and accelerated weathering.

Methods of extraction

Separation from the rock mass

By cutting action:	*Mechanical devices* (usually in combination):
o drilling-broaching	o jacking
o wire-saw	o levelling
o diamond disc	o hydraulic breaking
o catenaries.	o water-jet penetration.
By heat action:	*By explosives:*
o jet piercing	o detonation cord
o plasma	o black powder
o electrical rock fracture.	o Finnish method.

Removal from the quarry face, using:

o hoists	o levers
o cranes	o jacks
o forks and shoveldozers	o inflatable bags
o winches.	o wooden or metal rollers
	o steel balls.

To make V-shaped groove
score two parallel lines
12mm apart along block

Slabby layers

Cut deep 'V' groove between
lines to divide block

Where water is readily available, working areas should be cleaned hydraulically. This exposes joints and clears loose earth and rock materials which both restrict the planning of the cuts, and can cause wire-saws to jam.

In smaller quarries, the freely parting slabs are prised open with crowbars. In some difficult cases, the layers have to be wedged apart, then subdivided by drilling holes, inserting wedges and driving them in until the slab splits, as for example the extraction carried out at a quarry in the province of La Rioja, Argentina. The material there is generally described as an oolitic sandstone of Permian age and belongs to the Orcobola formation. Main production is in the 2 to 7 cm thickness range.

Slabby layers are the most convenient deposit for extraction and working, provided that the material meets the required standard and specifications. When such deposits are near the construction site, they have been proved to be the most economic building material available.

The plane of easiest splitting is sometimes called the clift, as distinct from the *direction* of easiest splitting, known as the rift. Wedges can be manipulated to make the best use of the natural properties of the stone. Generally, the most successful split is obtained when the weight is equal on both sides of the wedge, which should stand upright. The stone should be split into equal halves as far as possible. If one side is weaker or lighter, the blow will tend to run out towards the weaker side, causing an uneven break. The wedge should stand at least half its length in the hole, but never reach the bottom.

If a bedding plane runs close to the line to be split, the wedge holes should point slightly away. There should be no delay between the blows, so that the whole row of wedges works as one. Firm and rapid blows are important, although extra-hard blows will make the wedge fly out.

Wedge pits or grooves are generally cut with a hammer and chisel. Where no jack-hammers are available, holes can be made with a round steel bar about 1½ metres long sharpened to a chisel edge, known as a *barramina*, or with jumper drills. A silica sand sludge or other abrasive speeds up the drilling. The groove tapers to the bottom, with its end finer than the edge of the wedge. Hole spacing and depth are dependent on how easy it is to split the material, and vary from quarry to quarry. To guide the line of fracture, deeper holes may be drilled occasionally, especially in particularly tough or soft stones.

In 40 to 60 cm-thick slabby layers, a V-shaped groove, about 10 to 15 cm deep, is chased along the whole length of the block to be cut. The pit is lined with scales (long strips of tin) and small wedges are inserted between them. These are gently tapped with light sledge-hammers until the stone splits. Chasing may take an hour per metre, although the actual splitting may only last a few seconds. The groove has to be cut accurately and true, otherwise the cut will split outside the groove. This method is used where no jack-hammer is available or where the drilling of holes would take more time than chasing.

Formerly, wedges were the main tool used for splitting stone or detaching stone masses by their insertion in induced or natural cracks of joint cavities, in a similar manner to the parting of slabs. Plugs and feathers have now largely replaced wedges and require holes for their insertion instead of wedge pits. The steel plug is rather more slender than the wedge, and the taper on the feather and the plug correspond.

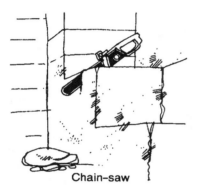

Chain-saw

Extraction devices With massive beds, heavy sledge-hammers are used in addition to wedges to subdivide the blocks, which are then cut and trimmed to the required size using smaller hammers of various shapes. Jacks are used to remove or displace particularly stubborn masses and, occasionally, explosives are resorted to, especially when the stone is obtained from building sites at the same time that foundations are being excavated.

The following notes are based on practices in

current use and briefly outline the nature of stone extraction operations utilizing the simpler devices.

Wedges

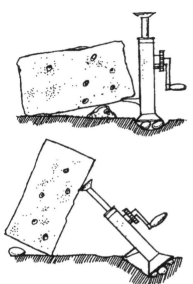

Mechanical jack

A set of wedges consists of three parts: two feathers or shims and one plug (see page 44). These are made to fit drill-holes of a specific size. The feathers form the outer part of the assembly and consist of two semicircular strips of iron tapered on the inside. The central part – the plug – is a steel wedge with straight, flat surfaces. The three parts are inserted in the hole as a unit. As the outer walls of the feathers are parallel, the force of the wedging is distributed over a larger area of stone than that possible with a wedge without feathers. However, wedges alone allow manipulations not possible with the more rigid feathers.

The thin edges (upper ends) of the two feathers are often finished off by small shoulders called lips, which facilitate handling and prevent the feathers slipping into the holes. Lips at right angles to the wedges are difficult to lift, and in practice they are made oblique and outward-curving, although this feature may strain and crush the edges of the holes. A slightly larger bore at the top of the hole should prevent this from happening.

Heavy sledging by hammer damages wedges and a technique involving light blows to give straight and even fractures is required. Flat surfaces on the equipment should be perfectly straight with uniform tapering, shaped in a swage or press-formed.

The purpose of wedging is to induce an even strain before the fracture takes place; the more gradual the strain the better the results. The time taken is critical; forcing the pace may cause irregular breaks. As different men strike with different degrees of force, the essential uniformity of strain and moderation in the fracturing rate is best obtained by one man. Gang-work, with a leader giving the strike cue, requires much practice but produces more rapid splitting. Plugs are sledged lightly in succession, beginning at

GUILLOTINES AND THEIR USES

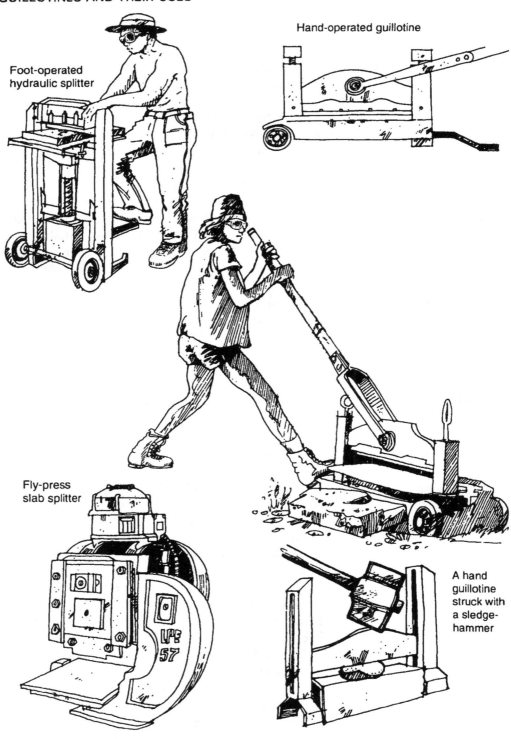

Foot-operated
hydraulic splitter

Hand-operated guillotine

Fly-press
slab splitter

A hand
guillotine
struck with
a sledge-
hammer

[52]

Various stone splits can be achieved by ......

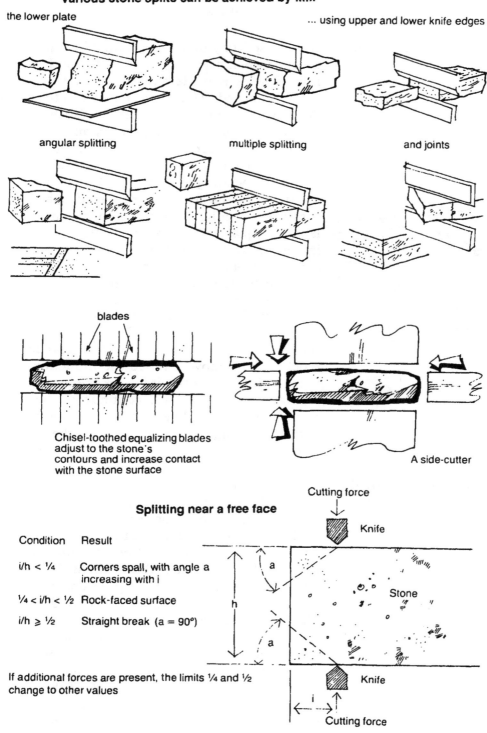

the lower plate

... using upper and lower knife edges

angular splitting

multiple splitting

and joints

blades

Chisel-toothed equalizing blades
adjust to the stone's
contours and increase contact
with the stone surface

A side-cutter

Splitting near a free face

Cutting force

Knife

Condition	Result
i/h < ¼	Corners spall, with angle a increasing with i
¼ < i/h < ½	Rock-faced surface
i/h ≥ ½	Straight break (a = 90°)

a

h

Stone

a

If additional forces are present, the limits ¼ and ½
change to other values

Knife

i

Cutting force

[53]

one end of the line to maintain an even strain gradient on the rock.

Block extraction Natural splitting directions should be taken into account when preparing a block for wedging. A row of holes is made along the line where the break is desired. The distance between the holes depends on the nature of the rock (such factors as hardness, size of grain, porosity and cementation), and the depth is determined by natural bedding planes, or economic considerations where no bedding is apparent. When the plug is driven in, the feathers are forced apart with pressures evenly distributed over their full length. Equal strain on the driven wedges is important and gradual fracturing will produce a uniform break. Alternate holes made to half depth distribute the strain more evenly.

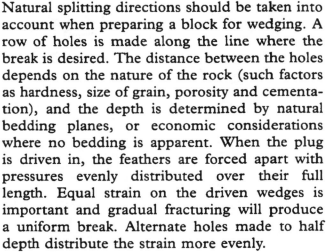
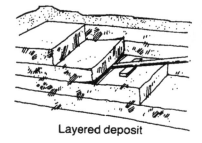
Layered deposit

The wedges are tightened in the drilled holes by hammering consecutively on alternate wedges with one or two light taps. The row of holes in the stone constitute a line of weakness in the block to be removed. Since the induction of a break or fracture causes it to seek the line of least resistance, the break will continue in the relatively low density line and split it along the combined wedge-web mass, leaving the wedges untouched.

Raising blocks from the quarry floor

When the block to be extracted (by undercutting, bed-lifting or floor-breaking) is freed on all sides by natural joints or vertically cut channels, horizontal holes are drilled at the base, some 20 cm apart and to a depth of not less than 20 cm from the back channel. All the channels have to be absolutely free from obstructions before a break is induced. Heavy, short wedges are placed alternately with long wedges; the latter should reach the maximum distance in the holes. Edges are colinear to produce the strain along the fracture plane. Maximum tensile strain is especially important at the back of the block. All wedges are tightened to the same tension and driven in completely from one end of the block to the other until fracturing commences. The short wedges are then driven in to make the break. A hammer weighing between 5 and 6 kg is used.

In some cases where no joint, vein or sheet structure exists, the artificial induction of sheeting plans may be appropriate. Holes are drilled to 2 or 3 metres and many successive small charges of black powder are set off at the bottom, to induce fracture planes. The sheet can be separated from the mass below by compressed air blown through pipes cemented in the holes, and granite layers are then peeled off.

Transport

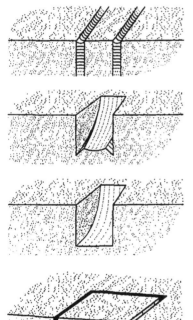

Pick channelling

In the beginning, man made his home where stone was available, using the material on the spot for various purposes. Later, when transport was required, human head, human back, camels, water buffalo, oxen, donkeys and mules were used, as they still are in many parts of the world. For limited demand, these means are less expensive than providing access roads.

Large stone blocks can be moved with levers, rollers and balls by pushing, or with pulling equipment, using ropes which can be manipulated with pulleys and capstans. The blocks can be moved on with sledges, cradles and wheeled carts. Ramps built from loose local materials like sand, gravel, earth or even bricks are additional aids, and manoeuvrability can be improved by using wooden uprights or cairns of stone.

Work in progress at a limestone quarry in India (above) *and a granite quarry in Quebec* (below).

Removing building stone from an underground quarry in India.

Site rehabilitation is essential and should be planned before extraction begins. Landscape near the highway to Jerusalem (Bab el Wad) was scarred by quarrying in the 1950s (above) and subsequently rehabilitated by afforestation in the 1990s (below).

5 Stone and the environment

With the world turnover of dimension stone increasing more than 30 times over the past 70 years and with the corresponding spread of sources, it has become more important than ever to be aware of the environmental impacts of quarrying the stone and its use. Quarrying is considered not as an end-use but only as a temporary utilization of land. An essential part of the planning process must therefore be consideration of environmental impacts both during the quarry's working life and after production has ended.

A reclamation and rehabilitation plan should be prepared at the same time as the extraction plan. Reclamation refers to the preparation of a site, including the creation of level surfaces or softening of quarry faces, while rehabilitation means returning the land to utility (which may be as building land or turning it into recreational, industrial or agricultural areas), though the terms tend to be used interchangeably. It is important that it should not be cosmetic treatment but integral to the quarrying process.

Apart from the need to plan the quarry sites and rehabilitate them after use, there are waste products which require attention. Enormous quantities of sludge are generated, which in marble and limestone production can reach 20–30 per cent of the weight of the stone worked; in the Apuanian marble-working regions this can be 2000 tonnes per day. The sludge can take oxygen from water, killing flora and fauna, and coastlines become dirty when the sludge is deposited by rivers. There is also potential pollution from silicon carbide used in grinding and from lead used in polishing, which are not biodegradable.

Of course, cement plants, brick kilns, and steel

plants all cause pollution and consume large amounts of energy in processing building materials, and in this context, the use of stone is well worth considering, particularly for any use within a 10-mile radius of its source so that transport costs do not outweigh the financial and environmental benefits.

Stone is the natural material *par excellence* which has withstood the test of time, and towns or villages built in stone have a clean look which is difficult to duplicate in manufactured materials. Stone is rot- and termite-proof, with low fire hazard and little likelihood of vermin contamination. In terms of the human environment, these are obviously desirable qualities.

Furthermore, in a desert, stone houses can be designed to contribute to water conservation on the dewpond principle, used by the ancients to gather water in desert areas. Proper utilization of the thermal properties of stone can induce energy savings by reducing the need for air conditioning, cooling and heating requirements, with commensurate decreases in pollution.

The effects of quarrying on the environment

In Chapter 4 there are some guidelines to the control and planning of stone extraction so as to minimize the negative effects on the environment; here the main factors are examined in more detail.

Waste control

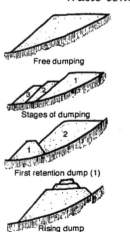

Free dumping

Stages of dumping

First retention dump (1)

Rising dump

The wastes from stone production can be controlled by using settling tanks, reservoirs and holding lagoons. The sludge acts as a filter for heavy metals and, after drying and separation of the harmful materials, forms an inert, non-polluting fill for roads, etc. – as does the waste stone itself. The wastes can become a marketable commodity instead of forming unsightly heaps.

A waste site should make good use of the topography, with the waste placed true to the landscape, in an area used only for waste disposal. It should not offend the eye or change the water regime, or cause undesirable seepages or soil creepage. Repository sites can include shore

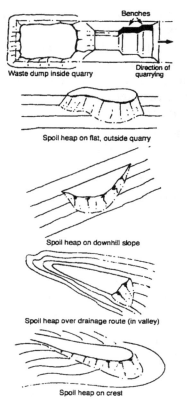

Benches

Waste dump inside quarry

Direction of quarrying

Spoil heap on flat, outside quarry

Spoil heap on downhill slope

Spoil heap over drainage route (in valley)

Spoil heap on crest

extensions, swamps, recreational development projects, jetties, dams, and road-building projects.

It has been shown that marble sludge can be used in the manufacture of prefabricated concrete products, for recycling lead from used batteries, and in agriculture. Brickmaking can use up to 40 per cent sludge.

The lime used for preventing rust during granite sawing can be neutralized by filter-pressing the sludge which will react with air and absorb carbon dioxide. The lead used in polishing accounts for less than 2 per cent of the sludge generated and apparently remains bound to the sludge.

Air pollution is controlled by using dust collectors on compressed air equipment to keep silica out of the atmosphere and the quarrymen's lungs. Thermal jets find less use in quarrying nowadays because of the noise, and those used during processing are in sound-proofed cabins with conveyor belts to pass the slabs under the jets. Gang-saws are increasingly located in sound-proof compartments and loaded by automatic trolleys.

Planning The aim of environmental planning must be 'maximum productivity with minimum damage'. Easily accessible stone resources are widely distributed throughout the world and in some countries stone is quantitively the largest mineral resource. It is widely realized now that some sites, where a large proportion of waste may be expected, can constitute unnecessary disruption of an ecosystem.

Location of quarries for optimum use of resources and limited damage requires regional monitoring and control. Relevant steps include keeping an inventory of stone resources on national, regional and local levels. Environmental considerations start at the prospecting level, as stone sources range from flat plateaux to high mountains, from river beds to coastal deposits, and from deep pits to shallow holes in the ground. Well-bedded and well-jointed solid deposits cause the fewest problems and allow a development plan with regular quarrying benches, although an apparently straightforward site can be complicated by tectonics, karstic features or other inherent characteristics. Slate deposits can cause problems, where extraction often follows steep inclines.

The rehabilitation potential of working sites, safety and energy-saving factors, markets and conversion facilities for waste and by-products (such as low-cost housing and road-paving projects) in the vicinity all need to be considered in advance.

Inventories are kept by the planning and licensing authorities which also keep land-use maps with geotechnical data. In many countries, the renewal of quarry operating licences is conditional on the fulfilment of a reclamation plan, an important principle being that the sites should remain in active use, in one way or another, after quarrying has ceased, and unproductive land returned to productive use. One possible statutory control involves the polluter paying according to the scale of the reclamation problem created, which can be a strong incentive to keep the need for rehabilitation to a minimum.

The land end-use should be decided before the extraction plan is prepared, and the waste repository sites considered. The cardinal rule is that the covering-up of pay material on future development areas is to be avoided.

Proper extraction methods have to be devised and applied, slope control established, safety features implemented, noise and dust controlled, and the undesirable effects of controlled blasting minimized. Attention to bedding directions and the size of the beds is of primary importance.

Access roads and haulage are to be designed to enable unrestricted development of the quarry, taking into account the proper design of slope angles and inclines, curves and banking. The road network is to be planned within the site and extended gradually to the processing plants which will need to be near the site because of the high cost of haulage.

Damage limitation   Optimum organization of stone extraction involves working in stages so that the visual impact is diminished and the obsolete parts are restored during the lifetime of the quarry. Ideally, dimension stone should be cut with equipment capable of providing regular geometrical benches and working faces rather than jagged or irregular contours and outlines. Minimization of unsightliness has been achieved to a large extent by the use of wire- and chain-sawing equipment rather than more costly pneumatic or hydraulic drills.

Controllable heights and widths of terraces are important and the resulting topographic form is to be conducive to vegetation growth. Any loose rock can then be within reach, making it easier to check landslides and rock falls, and quarries can be continued underground. This has the advantages of avoiding scars on hillsides, pollution of the atmosphere and noise while obviating the need to clear prohibitive overburden and creating useful space when quarrying is finished.

Rehabilitation and replication   Different treatments are possible, according to the quarry type: the landscape can be restored to the way it was before quarrying began (replication) or the area can be reused in some way. A few of the more common approaches are listed below.

Hillside quarry site: landscape sculpturing, camouflage netting, staining of quarry faces.

Pit quarries: landfill, fish farming, paddy fields, artificial lakes, parking areas, olive groves.

Underground quarries: storage, refrigeration, earthfill.

Hillside quarry rehabilitation is relatively simple, but flat areas are more difficult, especially where waste dumps are visible. Inevitably, it makes a difference whether dumps are inside or outside the quarry, on a downhill slope, over drainage routes or crests. Pit quarries tend to fill up with water, which affects end-use. The costs of rehabilitation must be considered at the planning stage as these can be very high.

Steps in rehabilitation include complete infill of slopes, backfilling slopes, selectively placing fill along lengths of the quarry face (especially in replication), blasting to fissure the near-vertical faces for natural colonization and vegetation.

Case study: Olival quarry

A good example of environmental planning is the Olival quarry at Villa Vicosa, in Portugal. For a few years in the 1970s, it produced white and cream marble and building stone from a site measuring about 40 000 square metres. When it closed, it left behind two large waste dumps, one at its northerly end and one at its centre.

In 1989 permission was given to reactivate the pit on condition that the excavated areas were reclaimed with contours and vegetation in keeping with the local surroundings. A budget and outline plan of the proposed extraction and rehabilitation phases was drawn up. Budgetary considerations included the cost of waste volume haulage to prospective dumps and replanting of reclaimed areas. Rehabilitation costs were found to represent between 1 and 3 per cent of the quarry's gross income. It was planned that after 57 years, the area would be rehabilitated to allow the traditional production of olives across the whole area.

Rehabilitation plan for Olival quarry

(Phase) No. of years	Central waste dump	Northern waste dump	New quarry A	New quarry B	Employment generated
(1) 2 years	Complete rehabilitation	Temporary improvements, topsoil, drainage and grass	Extraction of 6600 m³ of material	Surface soil removal	Quarry-related productive activities
(2) 14 years	Olive grove maintenance	Temporary slope correction, topsoil and grass	Extraction of 46 200 m³ of material	Surface soil removal	Quarry-related productive activities
(3) 41 years	Olive grove maintenance	Complete rehabilitation, olive tree plantation	Complete backfilling, waste dump formed from quarry B until B is exhausted. Final rehabilitation and olive tree plantation	Extraction of 136 500 m³, backfilled from northern dump and from waste dumped on A, rehabilitated and planted with olives	Quarry-related productive activities. Final rehabilitation with olive trees to return to traditional agriculture

In the first phase, lasting two years, two new areas of excavation within the site were proposed. At the same time, the central dump was immediately rehabilitated, slopes were stabilized, topsoil replaced, drainage systems dug and olive trees planted. The second dump was used to backfill one of the new quarry sites as it became exhausted. The table shows how the overall operation was planned.

Case study: Estremoz quarry

The Estremoz pit quarry is located in the same area of Portugal as the Olival quarry and it is one of the deepest quarry excavations in the area. The diagrams on page 66 show the stages which were used in rehabilitating parts of this quarry while allowing new work to proceed.

o Pit A was completely refilled with material from 1 and 2.
o Large-scale equipment was installed on the new flat area at A and a new quarry below 2 was opened where good quality marble had been found.
o Water was drained from pit B, a slope stability analysis was carried out and, using new techniques, marble extraction was restarted on the northern face of pit B.

By moving the material from 1 and 2 into A, the adverse visual impact of the spoil heaps was removed, a flat area was formed on which heavy equipment could be installed, and a new area of excavation at C(2) was made feasible.

Case study: Ontario

Until recently, Ontario has levied 2 cents per tonne of aggregate material extracted for the rehabilitation of land. This money was paid into a fund and was not returned to the operator until rehabilitation was completed. Unfortunately, many operators viewed this levy as just another tax and did not carry out rehabilitation work as the funds did not reflect the costs involved. The levy was increased to 8 cents per tonne in 1981 and began to take effect. Unfortunately, the notion of rehabilitation was deemed to mean the speedy restoration of land to any socially

1. *Estremoz quarry before rehabilitation had begun: A = abandoned pit; B = active pit; 1 and 2 are dumps*
2. *Material from dumps 1 and 2 put into pit A, ultimately creating a large flat surface on which to install a crane*
3. *New pit C established to extract good quality marble from the site of dump 2*

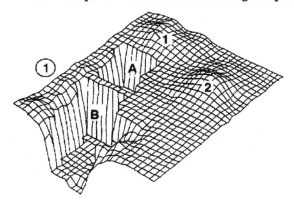

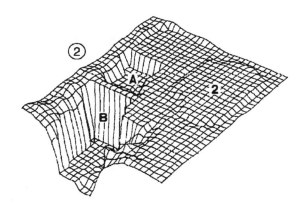

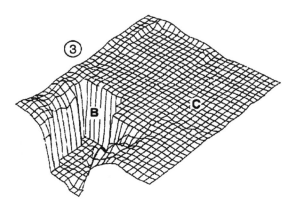

acceptable function, be it urban, rural or recreational. It is thought that over half the available funds for rehabilitation have been used in cosmetic but non-effective measures. Now the Ontario government is taking action, and several models are being considered.

Government financial security model

Each licensed property is levied as above, but with a minimum levy of $1000 and a maximum of $3000 per disturbed acre. An annual refund is paid to the operator in compensation for any rehabilitation work carried out that has been approved by a government inspector.

Standard rehabilitation model

This model, drawn up by McLellan (1984), provided a clear concept of quarrying as an interim activity, which divided a site into four distinct zones:

1. Land containing untouched reserves and still being used for its previous land-use activity
2. Land containing untouched reserves, but being prepared for extraction activity
3. Land containing the actual working mine
4. Land which, having been mined, had been returned to its previous land use through rehabilitation

Important in this model are the notions that mining is only an interim activity, that the process should be progressive and sequential, and that the minimum area of land should be disturbed at any one time.

Agricultural technical staging model

Another model, by MacIntosh and Mozuraitis (1982), is a refinement of the previous model incorporating a greater degree of pre-planning with reference to the quantity of overburden removed at any one time; the storage and handling of topsoil and subsoil separately; techniques to provide for a speedy return to agricultural use.

The major omission from all these models is the setting of goals with respect to the rehabilitation of land. If legislation is to be effective, an evaluation programme should be devised to assess the amount of disturbed land that should exist on any licensed property at any one time and the rate at which rehabilitation should take place.

Effects of the environment on stone

The effects of weathering differ, depending on the type of stone which is used. Homogeneous, light-coloured stones, which show variations more than textured sedimentary ones, should be avoided in situations where weathering and premature exfoliation are likely to occur.

Some light-coloured granites darken appreciably when wet, and this effect can be exacerbated by certain surface finishes which allow water absorption. Where rising damp is likely to be a problem, even if only for short periods of time during floods or snow, it is advisable to use relatively impervious dark-coloured stone for the lower courses of buildings.

If an area is prone to earthquakes, the correct stone must be selected. For example, granite, an immensely strong stone, may contain fissures and will therefore be unsuitable for foundations in earthquake-prone locations.

The effect of the bedding direction on a stone's durability cannot be overemphasized. Face bedding is especially prone to exfoliation and should be avoided. This should be considered even at the extraction stage, when the end use of the stone should determine the geometry of the extracted rocks. If stone is to be used as a thin veneer, small areas of weaker stone may be a problem if the point of fixing coincides with the weakness, and once again it is necessary to know the end use at the quarrying stage.

If the correct stone is used, maintenance is unlikely to be problematic. However, there are few remedies if the wrong stone is selected for a particular context, or if it is subjected to atmospheric damage. This fact has become more important as thin (10 mm) veneers, now used extensively, are less forgiving than solid stone used traditionally and will not accommodate cracks, spalls, fractures and other micro- and macro-blemishes in the same way as solid blocks.

Spontaneous reclamation of 'piedra adobe' volcanic tuff working by a small-scale quarry operator in Bulacan, the Philippines: the excavation in the background has been filled with soil and turned into a paddy field.

[69]

Dust collecting equipment (above) protects both the user and the environment in Minnesota, USA, while operators use appropriate safety wear – for example, a hard hat in the processing plant in Belo Horizonte (below left), where granite is shown being hand-polished, and ear protectors in a stone processing plant in Alicante, Spain (below right).

A well-benched breccia marble quarry at Peqiin in Galilee.

[71]

Large wagon drills are mounted on the saddles of quarry bars for accurate drilling at this marble quarry.

[72]

6 Tools for stoneworking

Terminology The terms used so far in this book have been predominantly geological or extractive, yet the same expressions used in architecture, masonry and construction may have quite a different meaning. The glossary should help. Since this book puts emphasis on intermediate technologies, words such as 'boning' or 'pig', which have traditional connotations, have been retained for want of better alternatives.

Tool names differ from country to country and even within regions, and usage also changes with time. Hence the duplication of some terms until an internationally recognized stone terminology can be agreed, preferably including terms used in intermediate technologies which might otherwise be neglected. For example, the four edges of a stone face are termed pitch lines or arrises. This is derived from a time when the edges were produced with a chisel. Nowadays, most ashlar is pre-sawn or split, and arrises might well be applied to any straight edge.

'Bed' to the mason means the horizontal surface on which stones are laid in mortar; to the quarryman, a plane of stratification; whereas for the geologist, it would refer to a continuous mass of sediments deposited underwater at the same time. A joint may be a space between stone units, or a tectonic feature in a quarry. Texture to an architect means a three-dimensional surface enrichment or ornamentation, but is two-dimensional to a petrographer.

Tools and techniques Stone can be worked by the inexperienced and the skilled alike. Much is learnt by experimenting, still more from mistakes, especially in the early stages. Basic actions include breaking, shaping, fragmentation, splitting and cleaving, bruising, flaking, abrading and polishing.

The principal action in the working of stone is

shaping, sizing and subdivision and the removal of surplus material. Dressing gives a special surface finish and is achieved by bruising and flaking, or by splitting actions performed by hammer or chisel. The stone is bruised into powder on the surface, the dust brushed off, and bruising is carried out again. Similarly, flaking and splitting are repeated to achieve the desired effect.

Earliest tools included stone hammers made from a stone harder than the workpiece, followed by copper and bronze tools, hardened by constant hammering. Many types of stone material can be worked with woodworking tools. In some parts of the world with woodworking traditions, it has never occurred to the craftsman to use his woodworking tools on suitable stones lying around his habitat. Yet the mortise and tenon techniques of the woodworker are present at Stonehenge and carpenters' joints can be seen at the Sancha Stupa in India.

Next in importance to the hammer, chisel and drill in stoneworking is the saw, whether in the form of a grubsaw or the handheld revolving cutting disc which has now largely replaced cutting blades.

Tools can be broadly divided into those used for extraction and those used for stoneworking, that is, cutting, shaping and generally forming the workpiece.

Extraction tools consist of the following:

Hand	*Machine*
levers	jack-hammers
jacks	saws
picks	wire-saws
hammers	chain-saws
chisels	mechanical chisels

Stoneworking tools include:

Hand	*Machine*
hammers	guillotines
chisels (including pneumatic)	saws
drills	planers or fraisers
saws	grinders
grinding stones	polishers
polishing powder	thermal and water jets

As previously stated, the fundamental actions in stoneworking involve impact from the hammer and a cutting edge from the chisel. These are sometimes combined in cutting-edged hammers. A natural development has been the mechanization of these machines and the stone-cutter or guillotine has been developed: this device can be simple or elaborate.

Breaking by hand includes the use of striking tools like hammers or mechanical devices such as levers and jacks. Crushing by hand is achieved with special types of hammer like bush-hammers; abrading uses space chisels and specialized equipment like planes. The action of polishing is not completely understood; it is achieved using different techniques. Drilling can be achieved by breaking, crushing, abrading or a combination of any of these actions.

As far as extraction is concerned, except for the use of hammers of various sizes, no effective impact hand-tool has yet been devised to replace compressor-operated jack-hammers or the more recent hydraulic chisels or hammers.

Splitting devices

Stone is brittle and a high stress concentration near its surface will cause spalling because of its non-homogeneous nature. Spalling is the basis of stone dressing and chipping; it requires comparatively little energy; it works by causing localized cracking and the energy released by loss of cohesion is used to enlarge the break.

Attrition and abrasion are part of stonecutting; sawing, for instance, and, to a certain extent, grinding and polishing. Compared to spalling, abrasion uses far more energy, and machinery design should be based on breaking rather than on abrasion.

Splitting devices are now available for rock-breaking rather than cutting. The machines for primary and secondary breaking are similar in terms of construction, differing only in size depending on the thickness of the workpiece. Splitters can be mechanical, hydraulic, lever-operated, cam, fly-press, and occasionally pneumatic.

These splitters can be worked by hand, hand and foot, motor or other power source. Stationary, mobile and portable models are available. The tool may have a single fixed blade, upper and lower blades, equalizing blades, chisel edges or polygonal steel rods. Some models may have side cutters for squaring. Forces exerted by these machines may be up to 200 tonnes, although the force required is much smaller. High force capabilities diminish fatigue in the cutting edges whose main purpose is to apply a high pressure and not to cut. In order that stone can be progressively reduced to sizes suitable for practical applications, preferably in modular units, a series of stone-breaking or splitting appliances of various sizes will be required.

A simple, highly effective device to trim extracted slabs is a mobile lever-operated cum cam-operated cutter. The slab is placed centrally between the upper and lower knife edges and the opening between them is adjusted to the thickness of the slab by two screw spindles. When the lever is in a vertical position, the distance between the upper knife and slab surface is 4 to 5 mm.

Early guillotines were principally laboratory rock-splitters. Attempts to mechanize slate-splitting resulted in the development of a number of devices. In the last three decades, splitting implements have appeared on the market for specific purposes rather than as part of a unified system. One refinement includes separate teeth that can match the contour of rippled and irregular surfaces and transmit equalized pressures. Another device induces a stress field which produces a directional split. Another machine makes it possible to obtain rock-faces of a conchoidal nature with exaggerated convexity. An experimental model of such a machine has given good results. This works on the Brazilian Test principle, creating a concentrated load that diminishes rapidly with increased depth.

In stone-splitting, a linear relationship exists between the applied force and the cleavage length, that is, a longer break will require a greater force. Torsional twisting to simulate a mason chiselling the stone is not essential as linearly aligned knives are adequate to provide the various rock-faced dressing effects.

To create an ideal break in natural stone, the top and bottom of the slabs have to be parallel to allow continuous contact with the blade edges. In practice, such slabs are not always easy to find, and a groove to house the guillotine blades can be pre-cut on one or both sides depending on the accuracy and shape of edge required.

On uneven surfaces – some slabby deposits have undulations – a shallow groove can be cut with a hand-powered or air-powered chisel, or with handheld power tools with abrasive or diamond discs of various diameters and profiles. An adequately deep groove can be obtained in one pass where the nature of the stone and its surface allow. It is imperative that the groove is straight so that the guillotine can make good contact with the blade edge.

Sizing sedimentary stones

Where available, a slabby limestone or sandstone should be selected to enable the use of a guillotine-type cutter, as this requires the lowest work input. However, any limestone deposit with a fairly well-developed jointing system can be used, especially if little overburden is present.

The stone is loosened from the deposit by levers, jacks or, if available, power equipment such as power shovels or dozers. The loosened stone blocks can be subdivided by sledgehammers or wedging, depending on their size and the equipment available.

Sledge-hammer

This is an important all-round tool, and a spalling or chisel-edge hammer is generally used, the weight depending on the size of stone to be cut. With some experience, fairly straight edges and sides can be obtained by carefully judging the cutting position relative to the stone edge. The stone can be subdivided into units and rough forms shaped by a bull-set or walling-hammer. The sledge-hammer can be used with a jumper where no power-drill is available. Masons have various preferences for hammer patterns, based mainly on traditional rather than technological considerations. A wide variety of hammer patterns should be available for experimentation during the training of stoneworkers or masons.

Wedging The breaking of bigger blocks of stone, which cannot be handled by a sledge-hammer or where a compressor is unavailable, can be achieved with power tools such as drills and hammers in conjunction with wedging units which may be short, long, hand or hydraulic, consisting of plugs and feathers, or just wedges. The stone block is split along straight lines and subdivided with the help of shorter wedges or by sledge-hammer. The procedure with the sledge- or bull-set hammer follows. The general principles of wedging are discussed under extraction on page 51.

Splitting The next step is subdivision by guillotine. Although specially designed for acting on parallel or approximately parallel faces to ensure maximum point contact, guillotines are versatile and can be adapted to a variety of situations. If one side of the slab is sloping relative to the other, it is important that the traces to be cut should be parallel. Where the surfaces are not smooth, a trace can be grooved, its depth depending on the quality and thickness of stone and the width to be removed.

In freestones, an auxiliary groove in any direction will produce a fairly straight cut. In stones with direction properties (common in sandstones, granites and many limestones), this groove should be made in a cleavage or joint direction.

There are various ways to make grooves; by chisel, by running the stone over a saw or disc table to ensure a straight bottom, by using a hand-saw or a handheld power tool with an abrasive disc. Where large quantities of stone have to be cut, a hydraulic splitter in conjunction with a fixed grooving device, such as the saw or disc table, will ensure speed and precision.

A more recent development has been the use of a *hydraulically operated chisel* or 'hydraulic hammer', mounted on a self-propelling track vehicle and with an action similar to that of the hand chisel: a trace can be outlined, then by gradual pressure a fracture induced which may be rather uneven unless there is a pronounced splitting direction in the plane of percussion. However, depending on the thickness of rock and burden to be removed, an increase in production level may be obtained with rapid subdivision, and the resulting stones can be passed through a guillotine for final shaping and finishing.

Hammers The hammer is the stoneworker's most important tool, and the ability to strike accurately with the minimum of effort is important. Much time and effort can be saved by shaping stone with a hammer, if no other tool is available. The choice of face or edge for the hammer, its weight and the right size of hammer handle are all equally important (see pages 80–81).

Terminology is important, too. A *mason's hammer* denotes a particular type, whereas a hammer much used by masons is the *lump-* or *club-hammer* which weighs about 1 kg. The specifications for hand-hammers (BS 876:1964) are particularly useful. The head should be made from medium carbon alloy steel, preferably forged by dropping a heavy weight on it, and heat-treated to minimize cracking and splintering. Stone hammers, such as the *cutter's hand-hammer*, can be obtained in graduations of 250 g. The weights for *stone-dressing hammers* range between 1 and 6 kg, and some sledge-hammers, which are used to break large rock-pieces, may weigh up to 10 kg.

Proper storage is crucial. High humidity may damage a wooden handle by swelling, whereas dryness will shrink it. The length of the handle is commensurate with weight and use: a sledge-hammer, usually 2 to 5 kg, is a double-faced striking tool and requires a straight handle with an elliptical rather than a circular cross-section, with a raised safety grip. The handle of a *stone-*

HAMMERS AND CHISELS

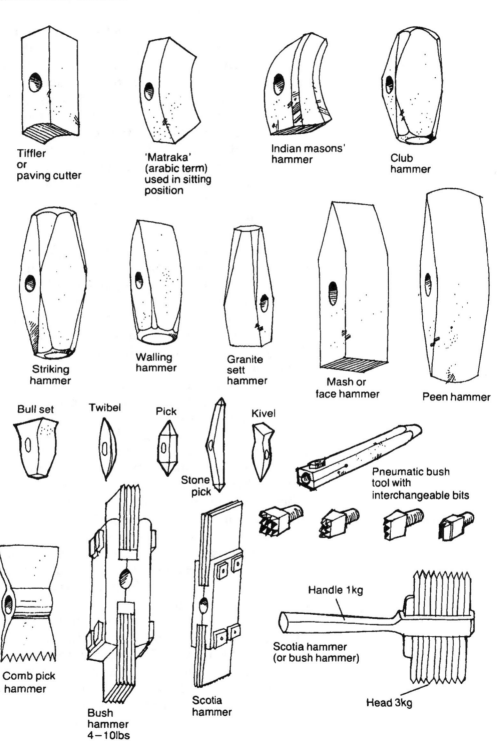

Tiffler
or
paving cutter

'Matraka'
(arabic term)
used in sitting
position

Indian masons'
hammer

Club
hammer

Striking
hammer

Walling
hammer

Granite
sett
hammer

Mash or
face hammer

Peen hammer

Bull set

Twibel

Pick

Stone
pick

Kivel

Pneumatic bush
tool with
interchangeable bits

Comb pick
hammer

Bush
hammer
4 – 10lbs

Scotia
hammer

Handle 1kg

Scotia hammer
(or bush hammer)

Head 3kg

[80]

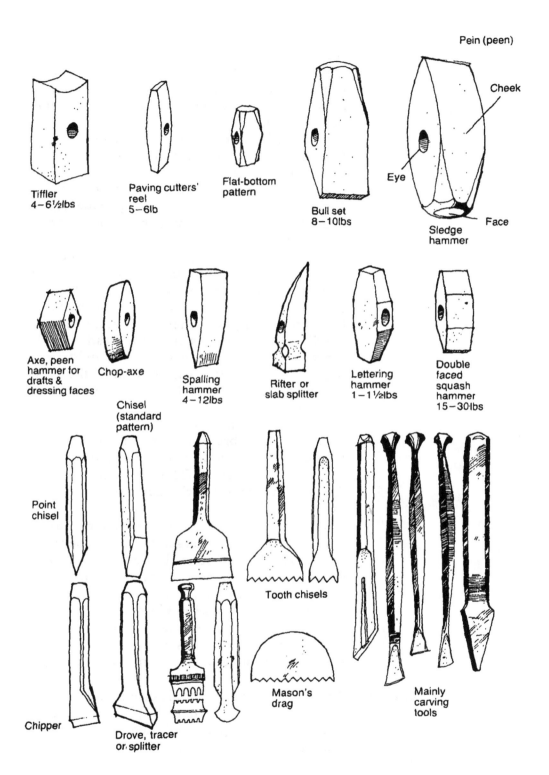

Tiffler
4 – 6½lbs

Paving cutters'
reel
5 – 6lb

Flat-bottom
pattern

Bull set
8 – 10lbs

Pein (peen)

Cheek

Eye

Face

Sledge
hammer

Axe, peen
hammer for
drafts &
dressing faces

Chop-axe

Chisel
(standard
pattern)

Spalling
hammer
4 – 12lbs

Rifter or
slab splitter

Lettering
hammer
1 – 1½lbs

Double
faced
squash
hammer
15 – 30lbs

Point
chisel

Tooth chisels

Chipper

Drove, tracer
or splitter

Mason's
drag

Mainly
carving
tools

breaking hammer, which is relatively long and slender, allows a very fast blow to be delivered. This rapid striking action has the advantage of producing considerable impact, overcoming the impact strength of individual pieces of stone being struck.

Wooden hammers, commonly known as *mallets*, are usually made of beech and ash, and are used for carving stone and where a delicate touch is required: working with claw chisels is for lighter finishing work. Hickory, or the nearest equivalent quality seasoned hardwood, should be used to make the handles of metal hammers.

The correct swinging of hammers, especially *sledge-hammers*, is important. Swinging a hammer from behind the body is a waste of energy and decreases control. A sledge-hammer should be gripped firmly at the heel of the shaft with one hand, and held loosely near the head with the other, then raised just above one shoulder and brought down sharply with full force, thus controlling direction and at the same time allowing a fast descent.

When the sledge-hammer is used for clearing isolated rocks, breaking can be facilitated by lighting a fire on top of the rock and throwing water on the hot ashes. This fire-setting tends to fracture the rock which can then be broken easily.

The sledge-hammer is particularly useful where slabs are too large, thick or uneven to pass through guillotine blades, and a fairly straight break can be obtained by moving the hammer back and forth along the line where the break is desired, and pounding it with a smaller hammer. A similar technique can be used to trim hard and irregular stones to size, where blows with a single hammer would slip off.

Chisels The chisel is next in importance to the hammer. Rudimentary chisels may be improvised from large nails; odd pieces of steel could provide well-balanced forged chisels. The size and width of the chisel required depend on the nature of the stone, especially its hardness and its end use. Tungsten carbide inserts prolong the life of the edge. Care has to be taken when using a chisel or hammer as excessive bruising may encourage subsequent exfoliation of the stone, although extensive or continuous decay is unlikely to result.

The *point* (or *punch*) is essential to rough-finish stone, and is usually 20 to 25 cm long. The punch is a heavier version of the point. The *stock* is hexagonal in section, and ground to a four-sided blunt (occasionally sharp) point. These chisels are used in the preliminary stages of smoothing rough stone, concentrating the hammering force to shatter the stone locally. The standard-edge chisel is similar to the cold chisel used for metalworking. A wider-edged version is the *bolster* with a bit width of 50 to 75 mm, which is used to chisel wide surfaces. The bit is short and a cutting level is ground on both sides. It can be used to smooth wide, flat surfaces or to split building stone, and it has the advantage that the wide blade transmits the energy of the hammer-blow laterally rather than at a simple point. A wider blade is termed a *tooler* and has a broader bit and heavier stock.

The *claw chisel* comes in various widths (25, 37, 50 mm) and teeth spacings. Holders are available to fit different types of claw edge. Its main use is to prepare flat surfaces by using claw edges of various fineness. It follows the point chisel to reduce the uneven surface into a series of shallow furrows for further finishing with the mason's chisel. The chisel is worked diagonally across the surface away from the edge. It is held at a 45° angle and driven with a wooden mallet – cuts are slightly overlapped. The final smoothing is performed with the *mason's chisel*, followed by grinding down by abrasive means.

In addition to the chisels which are mainly used for working stone surfaces, several types are available for functions such as edge preparation and fluting. The *pitcher*, *set* or *drafting tool* is like the bolster, but is used at a steep angle to remove excess material, especially sandstone. The waste is cut away gradually, 25 mm at a time from limestone, and 37 mm from (softer) sandstone, by holding the cutting edge parallel to the line at just less than right-angles to the stone face. The pitching chisel is also used for rock-facing or pitch-facing work, and the angle of holding the chisel depends on the degree of rock-facing required. High-quality work should show no tool marks on the finished surface of the stone. The pitcher does not cut but fractures the stone: for cutting a chisel and a *club-hammer* of $2\frac{1}{2}$ to 3 lbs is usually used.

The *plugging chisel* (or seam drill) has a long flat skewed bit, often fluted. It is used to cut into a stone surface, to fix plugs for example. A *cape* (or cross-cut) chisel has a deep wedge-shaped bit, narrower than the hexagonal stock with a slight flair at the tip, and is ground on the upper and lower faces to an angle of 60°.

Whenever a chisel is used, it should be gripped near the head to prevent wavering, and pressed firmly against the workpiece. The chisel point and *not* its head should be watched (see page 89).

The mason is watching the chisel point, not the head

Power chisels These are mainly pneumatic hand-tools consisting of a steel cylinder, with movable pistons or hammers worked by air pressure, connected to a compressed air supply by means of a hose. Various tools can be connected, including hammers, double-bladed cleaning up bush chisels, and plain, toothed and carving chisels. The main advantage of power tools is that blows can be delivered rapidly, which is especially suitable for trimming actions. Pneumatic tools are not considered satisfactory on the softer types of stone, as their action is too severe and insensitive for fine work, although power chisels can be used for roughing out the softer stones. The much faster action compensates for the noise and vibration, and the use of ear-muffs and safety goggles is advised.

Drills Hand-held drills are the precursors of the vast array of drilling power tools and machines available. Plug drills still have limited uses, as have

[84]

compressed-air-operated percussion drills, which have been extensively replaced by motorized equipment – whether electric or fuel operated. Drilling equipment in stoneworking is used for extraction at the quarry, for subdivision and shaping of blocks, and it is increasingly used to drill holes for the anchorage of cladding slabs.

Some of the larger rotary drills used for extraction can also be used in exploration; these are referred to on page 127. For extraction, compressor-operated jack-hammers, wagon-drills and jumbos with various rail or quarry bar arrangements are available. In many instances, these are being replaced by hydraulic equipment. Such equipment can be used for channelling, drilling holes to take wedges, providing conduits for wire-saws and, in larger quarries, to cut tunnels and adits.

Drill bits are available in a large range from Archimedes-spiralled bits used in the softer limestones to ultra-hard carbide and diamond-coated or diamond-set drills capable of penetrating the most resistant stone material.

Saws Hand-saws are still used in quarries supplying softish stones which harden on exposure. Saws can be held by one or two workers. The saws may be toothed or may have a slightly wavy section, such as the grub-saw. The grub-saw has a longish blade and is either fitted in a work-frame and worked by two men, or power-driven. Generally, saws for soft stone types have teeth, unlike hand-saws for hard stones. The hand-saw can be suspended on a frame or motorized as a chain-saw with tungsten carbide teeth.

Bars The basic crowbar is straight, between 1.5 m and 2 m in length, with the greatest diameter that can be conveniently held and forged to a chisel shape. The bar is used as a levering tool in cleaning overburden, extracting and moving rocks, splitting and positioning large stones. A pointed version is used to induce a break in slabby beds. For any levering action to be transmitted, the bar

Above: *Wedges are inserted into sandstone then split with sledge-hammers in Nepal.*

Below: *Splitting sandstone, first with a hammer and chisel* (left) *then with a sledge-hammer peen.*

Left: *Using a sledge-hammer to split sandstone at a quarry camp in Nepal.*
Below: *Using a jack to wedge out stone from a joint in a deposit, Godavari, Nepal.*

[87]

HAND-
CUTTING
A PLANE
SURFACE

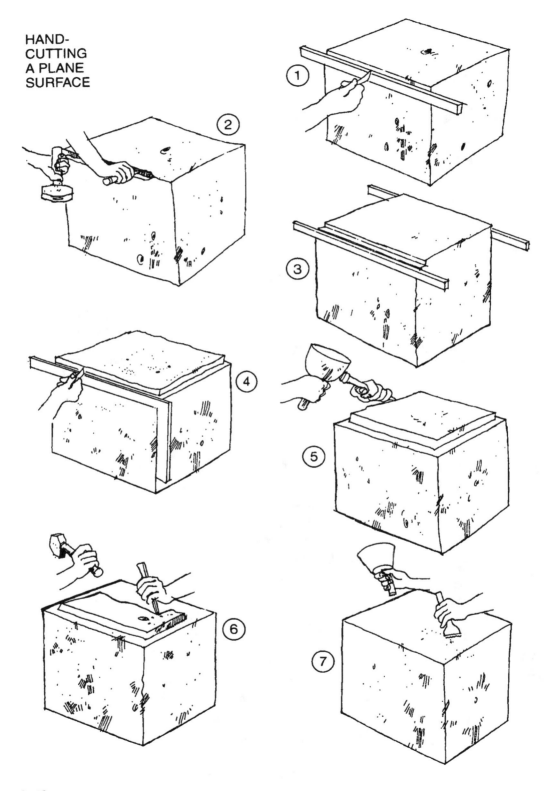

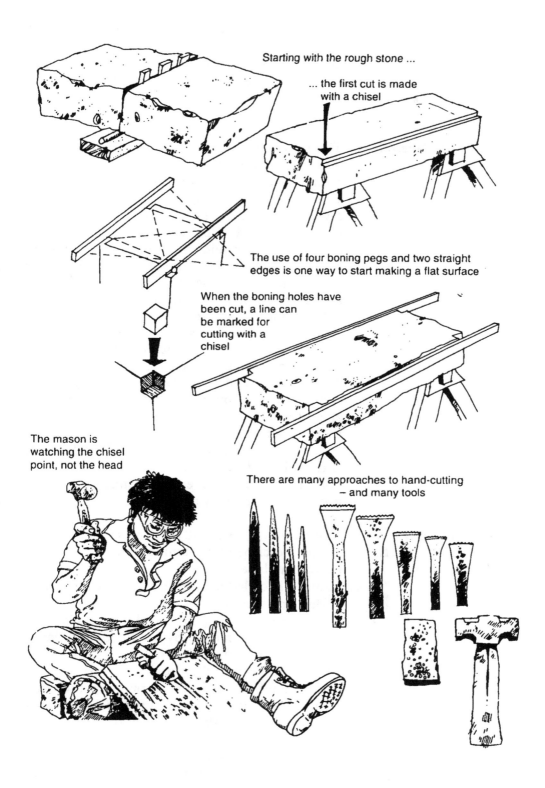

Starting with the rough stone ...

... the first cut is made with a chisel

The use of four boning pegs and two straight edges is one way to start making a flat surface

When the boning holes have been cut, a line can be marked for cutting with a chisel

The mason is watching the chisel point, not the head

There are many approaches to hand-cutting – and many tools

must rotate about a firm point (the fulcrum). The nearer the fulcrum is to the bottom of the joint or boulder, the greater the leverage, with the overall length of the crowbar limited by the quality of the steel. Crowbars or levers are the most-used tools in stone extraction.

Stone-breaking The procedures are similar to those in extraction, except that as the breaking is not *in situ*, it requires suitable support from benches, floors, sand-boxes, planks, pipes or angle iron. To subdivide a block higher than 40 to 80 cm, drilling may be required, depending on the properties of the stone. It should be determined from the beginning whether the material is required for ashlar (in which case dimension specifications are essential) or for non-ashlar constructions.

For stones measuring up to 40 cm a V-shaped groove is made using a two-headed pick-axe for softer stones and wedge pits for harder varieties. In both cases, wedges are used to make the break. Their shapes and sizes vary, depending on the stone, but it is important that in each operation all wedges are equal. The stone is then banked so that full attention can be given to shaping and dressing. In several countries, much of this work is done in a sitting position with the stone supported on the floor, either resting flat on the ground or diagonally supported.

Hand-cutting a plane surface Stoneworking, especially the mason's craft, can be well described by examining the steps needed to form a plane surface from a rough block, as it requires knowledge of both tools and materials.

Building stone, if not cut by machine, is supplied roughly shaped as large rocks and boulders, or obtained by subdividing a larger block. There is usually no level face or joint. To square a stone, the roughest protrusions are trimmed off by hitting the pitching tool hard against them and, using a punch held at a shallow angle against the stone, by making parallel cuts down the surface. To square one or more surfaces of the block, the first step is to straighten the upper face, opposite the face on which the block is standing. A line is drawn below the level of any surface depressions, to ensure a smoothly levelled surface (see page 88). An initial chisel cut is followed by another on the opposite side. Usually, such a plane is started by working a straight draft on the end of the block. With this method, any adjustment would require reworking a long draft each time. This can be prevented by using four small cubes of any hard wood, about 5 cm square, known as boning pegs. These are set at the four corners of the surface to be levelled and the straight edges of the cube are aligned by eye.

Above: *Splitting pre-sawn sandstone with wedges near Maseru, Lesotho. A shallow cut has been made with an electrical cutting tool to ensure a straight break.*

Below: *Trimming a block into building stones on a flywheel-operated guillotine in Cordoba, Argentina.*

[92]

Above: *Using a pick-axe to split stone in India.*

Below: *By using the natural planes of splitting in granite, short plugs and feathers are sufficient to subdivide granite blocks.*

Above: *Excellent example of a vaulted ceiling of ashlar.*
Below: *Stone abutment of a suspension bridge in Nepal.*

7 Architectural uses

The natural structure of stone, combined with careful processing, allows a finished product achieved with few other construction materials. Using stone as a building material can be costly or inexpensive depending on its application and the quality of workmanship. A major advantage is its versatility.

The structure and form of a building can be used as a means of expression by varying the use of the building stone to combine function with ornament. Monotonous expanses of wall can be relieved by adding a few courses of stone in different shades. An emphasis on horizontals can be used to highlight parts of a construction. Stone sills and lintels, while serving as integral construction elements, can also be used decoratively.

Types of building stone

The type of stone that can be worked depends on the availability of appropriate tools and the skill of the mason. Limited use is made of steel frameworks which require careful fitting; architects' drawings are usually considered sufficient, with attention to drafting and pattern-making where required. The main types of stone are ashlar, slabby, rubble and trimmings.

Ashlar

These are squared stones used for facing, dressed to provide various textures. These days, squared stones are prepared by mechanical means, reducing hand-dressing to a minimum. Basically, machine-cut ashlar is squared on five sides, with the back left rough if it is to be unexposed. The exposed face may be sawn-cut with a diamond saw, rubbed, rock-faced or pitch-faced, tooled, pointed, sand-blasted, brush-hammered and vermiculated or rusticated with various margins, usually of a pattern different to that of the central part. All these textures can be obtained by hand-dressing, mechanical-dressing or a combination of the two. The cutting of the basic stone by machine ensures squared and clean arrises and saves time in the final dressing, as the exposed face is regular and needs only the minimum of finishing. In hand-cutting, the ashlar is prepared either with hand-tools, rock-splitters or mechanical hammers and chisels.

The same types of dressing are used for both hand-cut and machine-cut ashlar.

Slabby stone This is stone produced in thin layers, used to face walls and fences. Thicker varieties are used in pavements, for example.

Rubble Rubble is shaped irregularly and the stones roughly trimmed to fit against each other in a wall. When the stone comes from fairly parallel, well-bedded layers, the natural bed is often used with no further treatment other than cleaning. Cleaning is especially important with a clay-like skin, which could weaken the mortar bond. Rubble may be found in many forms: polygonal, coursed, field-stone and random. It is often used in combination with ashlar as backing or filling between quoins.

Trimmings All regularly cut stone *not* in the form of ashlar can be classified as trimmings. This includes lintels, sills, caps, mouldings, coping, columns and entablatures (such as cornices, friezes and architraves), stone steps and arches. Pattern profiles made of thin zinc sheeting, plywood or other suitable materials should be used for making trimmings.

Types of dressing A description of dressing types is only of general value in written specifications. Tool widths, direction of application and pressure intensities vary from mason to mason. All the types of building stone discussed above may have one or more of the finishes described below. To save weight during transportation, the stones are roughly squared at the quarries.

[96]

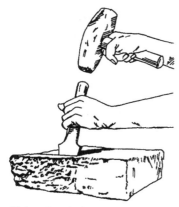

Using the pitching chisel for rock-faced finish

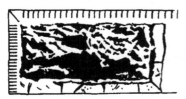

Rock-faced surface with chiselled margin

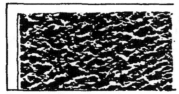

Rough-pointed surface with chiselled margin

Rock- or pitch-faced. The face is made to resemble the natural break as it comes from the quarry, and the joint edges are straightened. No tool marks are left on the stone, unless specially required.

Rough-jointed and fine-pointed. Projections on the face are taken off by using a stone-dressing hammer with a chisel point or pick, leaving a pock-marked dressing. This is also known as hand-picked, rough-picked, fine-picked or chisel-picked. The pick is used for finer work than the chisel, and the shallower and more delicate effect is produced by a short, direct stroke.

Drafted margin. A margin is cut a few centimetres from the edge, with a chisel, axe or peen-hammer. This type of dressing was used extensively by King Herod, for example in the construction of the Western Wall in Jerusalem, where rock-faced bosses project from the back-set margins, typically finished off by a serrated chisel dressing. In the larger stones, margins may be 7 to 13 cm wide and the bosses may project between 5 and 25 cm. The height of these stones usually exceeds 50 cm.

Tooled or drove. Parallel, vertical or horizontal lines of a regular pattern are cut and described as six cut, eight cut and so on, according to the number of grooves over a given distance – usually 25 mm. These are mainly made by machine.

Drag. Similar to tooled work, but the grooves are discontinuous. It is usually applied by a serrated chisel-edged hammer dragged over the surface. Short strokes are applied and the incisions measure between 25 and 50 mm in length for each stroke. The diagonal grooving typically found on buildings made by the crusaders was made by a similar tool.

Sand-sawed. This is the surface left on the stone as it comes from an abrasive-fed gang-saw.

Shot-sawed. Also known as a ripple surface, this is deeply scored or grooved using a steel-shot abrasive.

Furrowed surface

Diamond-sawed. This shallow arch-like groove is acquired by scoring with a diamond-toothed sawing disc.

Reticulated. Unusual dressing found mainly on older buildings. Irregular network of bands worked on a true-faced stone. The bands are sunk to about 10 mm deep and the face picked with a fine mallet-headed point.

Reticulated surface

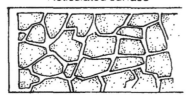

Rubbed. Finished by hand or more often by machine on sawn stones, giving a fairly smooth surface.

Bush-hammered. This is done by hand with a bush-hammer and has a rougher and more pitted appearance than similar tooled work. The hammer is faced with sharp pyramidical points, giving a dotted finish. The names used to describe the dot density include fine, medium and rough.

Vermiculated surface

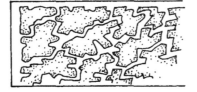

Vermiculated. Dressing with a worm-eaten appearance in the form of irregularly shaped sinkings and winding ridges.

Rusticated. Usually shown as fluting between courses and, less frequently, between joints, forming a sunken dressed margin.

Sand-blasted. A sand-blasted finish brings out the natural textures. The intensity of the effect can be varied by using different abrasive grainings and hardnesses.

Masonry practices

Although architectural and constructional details are only *referred* to in this book, the author's repeated observation of bad laying in stonework makes the following remarks and illustrations pertinent.

Bedding

The importance of bedding a stone cannot be over-emphasized. Rocks have been laid down by nature in a balanced manner and putting them down in any direction can gradually upset this careful stacking. For this reason stone should be laid down in the same way that it was deposited, on its bedding plane. The use of edge or joint bedding is less harmful if even-grained stones are used. Face bedding, however, is especially prone to exfoliation and should be avoided; an analogy is that of a book which, if laid flat, is unaffected by pressure from above, but which is damaged if it is placed upright and then loaded. There are some exceptions to this rule: in stones of uniform structure, for instance, the rule can be relaxed. The distinction between block masonry and slab masonry is essential when planning the use of a certain stone.

The thickness of the beds in the quarry is of primary importance and determines the way in which the stones should lie. The height of the mullions usually exceeds the thickness of the bed: hence, long slender mullions are placed edge-bedded, rather than using small pieces of stone piled precariously above each other. Edge bedding is also used for cornices, copings and string courses, to avoid face-bedding. To achieve a tall, slender masonry detail, the quarry bed may be cut diagonally across in order to obtain maximum height.

When masonry was carried out mainly by hand, the units were usually produced at the quarry site and attention to these details was taken for granted. Nowadays, stone may be fabricated many hundreds of miles away from the quarry site, sometimes to the detriment of the quality required.

Finishing The type of finish used depends on the nature and the texture of the stone. Generally, a hard, smooth stone can take any dressing. A hard, granular or calcareous stone should either be sawn or else given a rock-faced finish. With oolitic limestones, sawing and rubbing may spoil the natural texture; in a coarse-grained sandstone, quartz grains, being resistant to sawing or rubbing, are most likely to break away from the cementing material intact after sawing, producing a naturally porous effect.

Generally, the hand-tools used for dressing are more or less the same as the tools used centuries ago. The conclusion that these tools have been more or less continuously used was reached after careful comparison between present-day practices and masonry dressings found in excavations.

Some larger works, such as a medium-sized artisan's workshop, use pneumatic tools and machine-tools for shaping and tooling. The choice of the right tool is vital. The tools described in Chapter 6 give an idea of some of the hand-tools used in both quarrying and dressing operations. The use of the various patterns depends very much on local usage; the choice of hammer, for instance, depends on whether the mason works at a bench or in a sitting position. As stated previously, the importance of matching the right weight of hammer to the function is as important as choosing the correct size of hammer handle. The weight of the hammers used in dressing ranges between 1 and 6 kg. Hammering with a pneumatic tool is preferable to pounding with a hand-held hammer, which may bruise the entire surface.

The width of the chisel depends partly on the hardness of the stone. Tungsten carbide tips prolong the life of the edge. Good care must be taken during dressing, as bruising may encourage subsequent exfoliation, although excessive or continuous decay is unlikely to result. The application of too heavy a pressure during machine dressing may give an effect similar to a blunt chisel.

Building stones are often produced as a by-product during the extraction and processing of marble. Methods are quite similar in marble fabrication, except that more use is made of circular saws in cutting and also in several finishing processes to simulate hand dressing. For cutting, circular diamond-saws are now in common use. Coping or jointing with carborundum-saws is still occasionally practised, especially on the building site. A stream of water or recoverable lubricant cools the cutting edge of the saw and carries away the cutting. Carborundum or diamond-impregnated continuous-rim discs are used for jointing and other cutting arrises, and where a smooth finish is required. For general purposes, segmented diamond-sawing discs are used.

After shaping by machine, except where a sawn-surface finish is required, the rectangular stone is dressed by hand- or machine-tools, and except for the superior edges and joints, little difference can be seen between a machine- or hand-squared stone. The method of squaring by machine is an important time-saving factor and often reduces production costs, especially during the placing of the masonry on the building site.

Field-stone constructions

Field-stones were used locally as a building material long before architecture became a formal discipline. The term field-stone is arbitrarily applied to any loose stones, whether natural, shaped, worked or unworked; in fact any mix which is suitable or available. At present the use of field-stones in building construction is being encouraged in developing countries. In order to build low-cost shelters or walls, stone must be available locally. Sources have been described earlier, collection requires no special skill or equipment, and although splitting can be helpful, whole stones can be used. When stone has to be transported from riversides or shores, it weighs only half as much if it can take advantage of Archimedes' principle by being moved underwater.

The essential elements for field-stone construction are similar to those applied to processed stones and include lay-out, footings, height-width relationships, tying-in, flats for top stones, rectangulars for ends and corners, and longs for tying-in (binders).

The difference between a stone wall and a pile of stones can be seen in the laying process. Besides brittleness, which enables it to be broken with minimum effort, another important property of stone is its high relative density, which together with its rigidity allows it to be used to make strong, stable walls.

Removal of stone

When stones are prised out with a bar from layers on an inclined slope or at a steep angle,

The first step is to gather field-stones ...

... then sort them into shapes and sizes ...

... split rounded stones can be put together to make rectangular block shapes

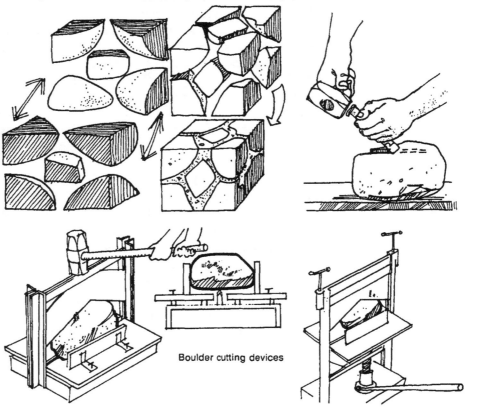

Boulder cutting devices

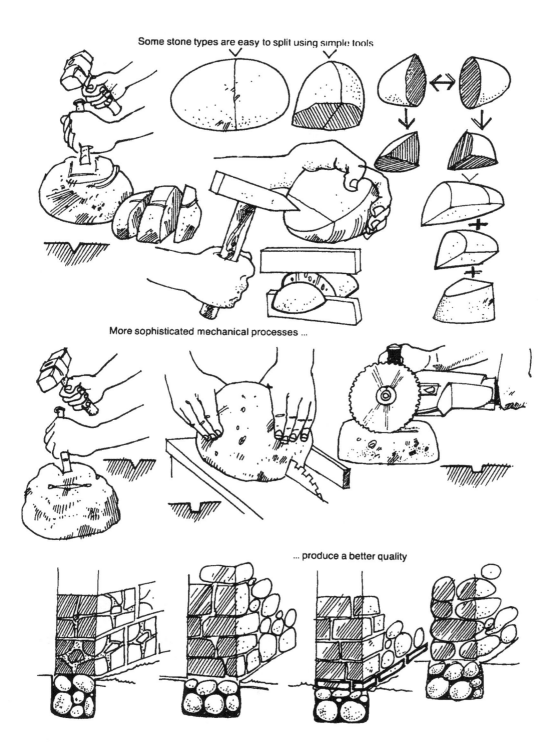

Some stone types are easy to split using simple tools

More sophisticated mechanical processes ...

... produce a better quality

loose stones should be thrown into the gap created to prevent the stone falling back. It is important to clear the surface of any deposit of stone about to be removed. The correct positioning of a crowbar or jack can sometimes bring tons of earth down a cliff.

Stone handling Stone handling is just as important as handling equipment, in ensuring safety, preventing tiredness and double-handling, preserving material and, last but not least, good housekeeping.

Stones, especially those that are very heavy, need not be carried. Let the weight of the stone assist rather than counter the move. The stone can be tipped up on its end or side, straddled or rolled between the legs. When lifting, only the legs and arms must be used, keeping the back straight and the stone close to the body.

Where larger stones weighing over 100 kg are being moved regularly, a hoist is indispensable. A block and tackle for heavier weights requires shear legs and gantries for support. Nylon straps about 6 cm wide are used to support the stone. Lewis bolts require drilling but can be used appropriately where such expanding bolts are available.

Building a wall A properly constructed dry-stone wall expands and contracts with changes in temperature, unlike a mortared wall which, being rigid, breaks and cracks under stress. The lowest courses, known as footings, should rest on subsoil rather than directly on the ground. The subsoil should be raked, where possible, to ensure a flattish base. Stakes higher than the planned wall are hammered in to mark the four corners of the wall. Cord is tied over the top of the stakes to mark the outline. Four more stakes are placed within this area, with the string touching the inner facing of each stake, and fastened to prevent curving by the wind.

The choice of stones for the foundation – whether to use small odd-shaped stones and save the biggest flat stones for the top course – depends on the nature of the foundation; soft, irregular footings may provide better anchorage than a flat underside. Open spaces filled with smaller stones provide good drainage especially when used below ground level. Angling the stones slightly inwards towards the centre causes the walls of the structure to lean in against themselves, increasing stability.

Another feature to be remembered during sorting is that stone sizes should be in proportion to the dimensions of the wall. Low walls require small and narrow stones, larger stones being set aside for higher walls.

Retaining walls There is perhaps no other type of walling in which stone is used as frequently as in retaining walls. Generally dry or mortared, there are several types of retaining wall; some basic structures are illustrated on page 107 and described below.

Vertical rectangular section which, though effective, is not economical. Because pressure decreases with the height up the wall, it is advantageous to have the greatest width of the wall at the base.

Vertical faced wall with sloping back. The increased area of the base gives more stability and a greater area over which to distribute the pressure. An additional advantage is that the weight of earth resting on the sloping wall side acts with the weight of the wall.

Section with a sloping back and battered face. This offers no practical advantages over the previous example, except that the battered face gives the appearance of greater stability.

Curved battered face. There is no advantage of a curved batter over a straight one, but it is considerably more expensive.

Sloping rectangular section. This section is wasteful, as the portion of the wall on the earth side of the vertical line drawn from the inside corner of the base does not act with the wall.

Although different principles may apply to dry walling, where batter is essential for stones which

WALL CONSTRUCTION

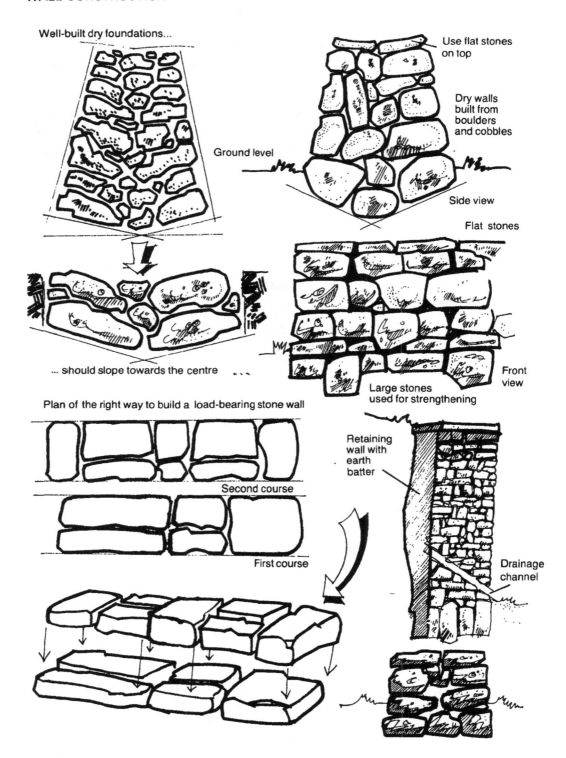

Well-built dry foundations...

Ground level

Use flat stones on top

Dry walls built from boulders and cobbles

Side view

... should slope towards the centre

Flat stones

Front view

Large stones used for strengthening

Plan of the right way to build a load-bearing stone wall

Second course

First course

Retaining wall with earth batter

Drainage channel

[106]

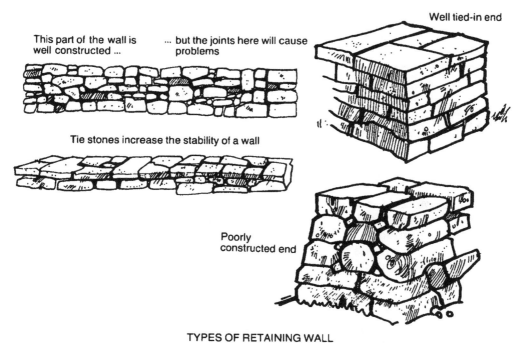

This part of the wall is
well constructed ...

... but the joints here will cause
problems

Well tied-in end

Tie stones increase the stability of a wall

Poorly
constructed end

TYPES OF RETAINING WALL

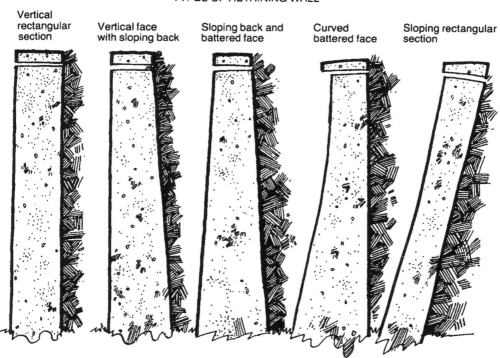

Vertical
rectangular
section

Vertical face
with sloping back

Sloping back and
battered face

Curved
battered face

Sloping rectangular
section

are not flat, with well-layered stones the moment of resistance (density × volume × leverage) is important and depends on the form and disposition of the wall.

The leverage is measured from the distance of

Building a wall with field-stones
Twenty points to remember

1. The best flat face of the narrow dimension of each stone should be facing outwards.
2. Each stone should be bedded solidly on the stones below.
3. A protrusion should be clipped off a stone, rather than trimming the stone by hitting it with small rocks and wedges.
4. Small stones used in the outer faces may work loose in time.
5. The thickest and heaviest rocks should be kept for the lower courses to avoid unnecessary lifting.
6. Joints between stones should not extend from course to course to form a run, or else the wall will fold, as dynamic stresses concentrate at the weakest point.
7. Every 2 to 3 m, a long tie-stone should be put in to tie one face to another.
8. Rectangular stones of even thickness, the longer the better, should be saved for ends and corners, especially outside edges.
9. The higher the wall, the wider and deeper the footings.
10. The top face of each stone should have a slight downward slant on which the next course can rest.
11. Splinters and wedge-shaped stones should be added to the main stones to keep them stable.
12. Wedge-shaped stones should be placed with the widest end in the core of the wall and the thin edge pointing out.
13. If built on an incline, a ditch on the upper slope of the wall and also a hole or culvert in the first above-ground course will permit drainage.
14. Batter is required for retaining walls, especially for the backward slope, the angle increasing with the height required. A rubble filling between the wall and the earth bank will enable water to flow freely behind the wall and prevent a build-up of pressure at any point.
15. A lattice fence should consist of alternate squarish blocks and thin slabs, with a support block where the slabs in the course above meet.
16. Gaps can be built at an angle to keep out cattle.
17. Stiles should be planned beforehand with rises of about 20 cm and the same distance forward per step.
18. The bottom of the stakes should be buried well below ground level.
19. Gates or door hangers should be fixed into the facing ends of the opening by drilling holes with a hammer and chisel, preferably with the stone in a horizontal position.
20. Stones should be dropped into place rather than put down so that they will find their own level.

---

a vertical line through the centre of gravity of the wall and the likely point of failure, usually at the intersection of the ground line and the back of the structure (see diagram on page 107).

Traditionally, dry walling is less common in the southern hemisphere. The composition of mortar is not as critical as it is in the northern climate, where it is subjected both to a range of temperatures and to frost. Usually, dry walling should not exceed 90 cm (waist height); even then, a base of 50 to 60 cm is required. However, higher walls can be built, widening the base accordingly (20 cm for each 30 cm of additional height), and most importantly, the footing depth should be well below any sensitive (such as frost level) ground effects, at about 60 cm.

Construction of dwellings

No structure is better than the quality of its substructure. This consists of the subsoil and the foundation laid for the building. Before starting construction, it is important to carry out a careful examination of the subsoil; that it is not liable to differential settlement and that the groundwater level will not affect its stability. The presence of back-filling has to be checked out and may require pre-consolidation. Last but not least, the performance of foundations already in the neighbourhood should be examined.

The walls of buildings resting on ground of variable strength often fracture, due to unequal settlement. To prevent this occurrence, the base of the walls is usually extended and supported by suitable foundations.

Foundations

The selection and the depth of the seating for the foundations depends partly on the firmness of the soil and the need to protect the seating from harmful climatic effects due to variations in seasonal humidity.

The object of foundations is to distribute the weight of the structure equally over the substratum and to prevent unequal settlement. The bases of structures are invariably made wider than the superincumbent mass, to increase stability by distributing the load over an area sufficiently large to withstand safely the pressure of the building and to counteract the damaging forces that tend to cause failure. These forces are described below.

The principal causes of failure are those which induce settlement: inequalities of earth resistance; lateral escape of soft soil; sliding of the substratum on sloping ground; shrinkage due to the withdrawal of water; atmospheric action; distributive lateral pressures causing overturn, such as wind pressure, the thrust of barrel vaulting, or an untied couple raftered roof; and lastly, concentrated lateral pressure which induces settlement and overturn, such as the thrust of framed floors, trussed roofs and groined vaults, which subject small areas of support to great pressures. Inequality of settlement in foundations has two causes: the unequal resistance of the soil and the unequal loading of the substratum.

With the exception of solid rock and gravel, nearly all soils are compressible under pressure and it is not always possible to avoid settlement. This is no problem provided that the settlement is uniform and of no great depth, and that the relative position of the parts of the structure remain unaltered. But where the resistance of the soil is not uniform throughout, there is a risk of irregular settlement. Buildings with irregular masses erected on uniformly yielding soils are subject to unequal settlement, too. Special precautions must be taken in both cases to distribute the pressure over a sufficient bearing area, or by piling.

When stone constructions are built on foundations which do not reach the bedrock across the whole structure, the foundations should be wider than the walls which they support. On stony ground, the building is usually constructed on or against the rock after it has been ascertained that the rock is sound, and *in situ*, otherwise the base has to be treated like a soil foundation. If the ground is sound and the rock uniform, then the lowest course can act as an initial foundation, but should then consist of cut blocks, the larger the better, for example 1 m long and 30 cm thick. However, since stone building is most economical using locally quarried stone, the construction should be designed to take advantage of the sizes available. Generally there are three types of foundation:

Foundations in trenches are 30 to 80 cm deep and wider than the wall planned. These are sufficient where the unit load is less than the practical resistance of the soil.

Foundations with footings. These are wider trench foundations, so that the load is spread over a greater area.

Foundations in trenches where the soil is problematic. In these cases steps or tiers are required.
Besides these basic principles, it is difficult to lay down hard and fast rules. In some parts of Africa where splashing rains may cause trenches to cave in, flat stones or boulders are required to protect against the erosion of the foundations. Generally, foundations are constructed by digging a trench in the shape of the house, tamping it down, then throwing or pouring concrete together with any stone material available in the area. Strength can be increased by selecting and stacking the stone with a minimum of concrete or mortar and applying the principles of dry-wall building (see page 106). The nature of the base of the trench determines whether concrete has to be poured, bearing in mind the weight of the incumbent structure and other environmental factors. For a compacted base of clayey soil or clay, with or without stone, a footing course may be sufficient for a one or two-storeyed building. For sandy, marly and other light soils, a concrete belt of about 20 cm is advisable, wider than the wall to be built on it.

Stone courses below ground level require as much attention as visible courses, since the substructure is more important than the rest of the construction as it determines the safe height of the building and carries more weight than any other course.

To complete the foundation the top of the

foundation-wall is levelled with long stone slabs, about 30 cm thick. As such slabs are not always available, or may be complicated to square with basic tools (as few hand guillotines can tackle lengths of more than 70 cm), a continuous concrete slab can be cast, on to which the first stone courses can be laid. A damp-proof course is always recommended, but not always practised. Besides using standard impregnated or plastic materials, damp-proof courses can be made of two or three thin slate courses, where this material is available.

Protection against running water is important, too, and drainage channels around the house, filled with pebbles or gravel, are usually effective. Precautionary measures against earthquakes are not only confined to the foundations, although these are an important part of the box structure. Basic precautions should be taken in earthquake zones:

o Stick to box features as much as possible. For example, internal walls should be continuous with supports at both ends.
o Use pilasters in corners and at doors.
o Roof supports should be divided over the outer walls – preferably a hip-roof.
o Balance the house so that for each opening in one wall, an opening is built on the opposite wall. This principle is based on one of the most important axioms for stoneworking and stonework, that a fracture will always follow the line of least resistance such as the weakest point or part.
o Foundations should preferably be continuous on flat ground. In mountains, excavation of the slope at the back of the house to form a terrace is preferable to building on fill, however well supported or contained, especially when the slope is steep, when sliding of the foundations will take place.
o Two separate rectangular or square houses are preferable to a single L-shaped structure.
o Porches should be continuous along one or more parallel sides of a house. If porches

have to be partial, then they should be in the middle of a wall to ensure some symmetry.

o Proportions of a structure are important – walls too high in proportion to the size of a house are earthquake-prone and one-storey buildings are preferable. The same applies to structures with very long walls, where support from an interior wall or from buttresses is essential.

o Doors, windows and other openings should be a minimum of 1 m from the corner. If too close the corners are weakened.

o Awnings and any other protrusions should preferably be part of the roof, with trusses being supported by the outer walls. An inner or central wall may be an integral part of the structure, but should not be a main support. Therefore, a roof should not be dependent on the gable, but its support should be distributed over the four outer walls; preferably the roof should be four-sided and of the hip type.

Walls Construction with stone is most likely to be associated with walling – even before the day that Joshua won the battle of Jericho and the walls came tumbling down! There are two types of wall: retaining and free-standing. *Retaining walls* are described in detail on pages 105–9. They are used to keep back earth, soil, fill, terracing or to preserve contour levels, to act as road-cuttings or control erosion; in all of these uses, one face is exposed. When built dry, such a wall does not usually exceed a height of 1 m and a thickness of about 45 to 50 cm. Retaining walls are usually started some 30 cm from the site and then back-filled.

Free-standing walls, such as a stone dividing wall, have both faces exposed. When dealing with ashlar, slabby or squared stones, the construction of the first course is fairly straightforward, but random rubble requires special techniques.

Following the foundation course or concrete slabs, the footing is set in a bed of mortar which has been levelled. Stones with vertical faces on

Simple building

the outside are set at fairly regular intervals to form a framework for the wall, and to mark the courses. It is important to use big stones for this to prevent settling problems which occur when larger stones rest on smaller stones, especially in irregular forms found, for instance, when random rubble is used. For solid walls, the work can be conducted from one side of the wall. If the two faces are constructed of stone cladding with fill in between, one face should be worked at a time. The procedure of big stones is then repeated, employing the principle of two-over-one and one-over-two, thus preventing lines of weakness and making sure that the stone slopes inwards to pull in the wall more tightly (see page 106). When filling in between two faces, it should be remembered that in-fill is a packing material, and not only does it not blind the wall, it also contributes to pushing the faces apart. The top faces should be bonded together properly with long, narrow stones, set aside during sorting. The topping requires a stronger mortar to ensure that the stones remain in place.

There are several types of *load-bearing stone wall*. The uncoated single wall is a simple structure consisting solely of the stone material used to build it. There is no means of stopping water permeating through to the inside face, as there is no impervious facing on the outer surface and no damp-proof membrane in the thickness of the wall, unless a damp-proof course is provided. The coated single wall does not usually have an impervious facing on the outside, but across the width of the wall a continuous anti-capillary damp-proof course is essential. If the internal wall is cladded, an air space at least 2 cm wide should be left, with wedges or battens used to maintain this gap. As regards insulation, apart from the commercially available materials (glass fibre, polystyrene and so on), it is possible to achieve the same results with, for example, wood shavings, dried vegetation or moss.

Cavity walls are external double walls consisting of two distinctly separate wall structures, more or less of the same thickness; water is prevented from permeating through the wall by a continuous air gap of 5 to 15 cm. The outside stone wall is usually non-load-bearing, but it must be resistant to external pressures, such as wind. This type of wall, referred to as a self-supporting facing wall, is usually built with cut ashlar or squared stone. Expansion joints are required for larger structures. This wall either extends to each floor level or is continuous for a number of floors; it should not be in contact with areas from which heat or dampness may be conducted.

Mechanical stability is ensured by connecting the exterior wall to the load-bearing wall in the form of ties made in non-corrosive metal and spaced at appropriate intervals. The thickness of each wall varies from 7.5 to 10 cm. The application of ties is made easier if the stone blocks used for the outer walls have the same dimensions as those used for the inner walls, so that their courses correspond. The blocks are squared or pre-cut to a height of up to 20 cm; their weight is less than 35 kg and joints of 5 to 15 mm are used. Movements between the two walls have to be considered and flexible joints must be provided across the air gap. Any roof projection of less than 30 cm is harmful and increases the amount of water reaching the façade immediately below; on the other hand, small projections at a lower level such as below windows are very effective.

As discussed under masonry on page 98, stones are laid in their natural on-bed position. In the case of soft, medium and hard stones, the underside of lintels, arches and some types of projection can be subject to attack by atmospheric elements, such as rain, in which case the stone is placed perpendicular to the bedding plane.

Composite walls may contain several materials across their width; for example, stone and concrete. The banked stone system involves pouring concrete between rough or cut blocks or slabs of hard stone which act as shuttering on the outer faces.

Composite walls may also be built between

two facings of cut or crushed blocks, filled with various stone materials such as pebbles, broken tiles, and a sand and lime mortar.

Stability of walls Walls tend to support one another, and a minimum thickness is required for their stability. The nearer walls are to each other, the greater the strength of the connecting wall. Walls enclosing a space give mutual support at their ends. Their thickness should increase with length.

Various tying and bonding devices compensate for inadequacies in stonework, serving to increase stress resistance and to impart a certain amount of robustness. These devices include corner ties, jambs or piers made of cut stone, hard-stone cupola ties, and iron or wooden ties which bind the corners of a building. Floors also act as braces or stays in a structure; as can partition walls, corner returns and buttresses.

Mortar The decision as to whether to construct a dry wall or, if mortar is to be used, then how much, depends on many factors, including the kind of stone available and the type and size of the structure required. The following points should be considered. Stone is usually stronger and more durable than mortar. Well-tied and stacked stonework without mortar is more stable, and often more attractive, than a wall held together with plenty of mortar. A major advantage of mortar is that, with its use, walls have to be less massive as it acts as bonding. Mortar seals the joints between stones, enables the use of different shapes and sizes and facilitates the use of rendering. Hard mortars are vulnerable to stress in walls. Mortar for stonework is less wet than mixtures used for other purposes to prevent dribbles and grime on the finished wall.

There are several different styles of pointing, but it should be noted that raised pointing should be discouraged as it tends to cover rather than highlight the stonework.

Stone as a roof covering Apart from slates, schistose, fissile and other types of cleavable rocks are used for roofing.

Above: *Limestone arch and other remains of Chastel Pelerin, built by crusaders in the thirteenth century.*

Below (left): *Limestone wall of roughly squared field-stones.*
Below (right): *The strength of the wall depends on both stone and mortar; if the mortar is too weak, the wall will fall down.*

Above: *Typical dwelling in the old city, Jerusalem. The joints between the stones of the limestone roof are grouted with a waterproof asphalt.*

Below: *Dry-wall shelter made from granite boulders and wadi stones, Wadi Faroun, Sinai.*

Roof coverings are generally between 1 and 2 cm thick and up to 50 cm square, as these dimensions provide durability and a warm and pleasing effect. Porous rocks should be avoided, as water absorption will add to the weight. Ceramic roofing tiles, unlike stone materials, absorb about 10 per cent of their weight of water. Concrete roof tiles increase roof loading by a factor of 2 and their life expectancy is a quarter that of slate.

The first step is to trim the material to various sizes, then sort out the lengths, as all slabs laid in the same course should be of the same length and, where possible, of the same width. The largest slabs are placed in the courses nearest the eaves, and the courses usually reduce in slab size as they approach the ridge.

Stone slab roofs have been reported in Italy (Piedmonte and Aosta region), Great Britain, especially in Oxfordshire and Gloucestershire, Silgharidoti in Western Nepal and other parts of the world. The laying of stone tiles is similar to slating, usually on battens which are fixed transversely on to common rafters. It is important to determine first the necessary diminishing gauge by the size of the material available. The spaces between the battens are then lathed and filled in with ordinary lime mortar brought up flush with the battens. Next, the slabs are bedded on stone lime mortar and pointed with the same material, then fixed into position by nailing with one galvanized iron nail near the head.

The slabs are cut and mitred at the hips and finished in a manner similar to ordinary slating, or covered with sawn stone or tile coverings.

Arches To construct an arch, a suitable temporary support is required which is traditionally made from wood and other fibrous materials such as bamboo or palm leaf; alternatively, a stone or brick frame or form can be used as a support during building. The construction of the arch itself is quite simple in practice if certain rules are followed.

A lintel is a straight arch. Single lintels over doorways and windows are more common in modern building than the triangular arch or

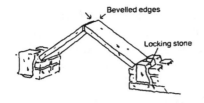

Bevelled double lintel

double lintel, which has been reported from Tigré houses in Ethiopia, for example. The double lintel is actually a peaked arch consisting of two matching slabs which are set against locking stones. The edges at the peak can be bevelled or locked with keystones which are inserted after positioning the slabs. Where sufficiently large slabs are not available, several suitably shaped stones may be used. The use of evenly shaped stone on both sides of the keystone is important to ensure balance and symmetry. The courses above peaked arches are normally built horizontally with special attention to their tying into the wall.

The double lintel lies between the straight arch and the curved arch. The curved arch is the strongest form of arch; it can be built from shaped random rubble as long as some form of symmetry is ensured. Taking the ultimate, perfectly cut stones can be used without mortar, as stone is strongest when compressed. Where slabby stones are too thin for shaping, they may still be used to form an arch.

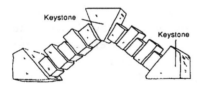

Peaked arch made of shaped stones

Corbelled stone arches can be constructed without forms but have restricted spans. A corbel is a projection out from the wall and, by continuously overlapping them, an arch effect can be produced. However, a good supply of long stones is required, as much of the length is used to cantilever the stone into the wall where the incumbent weight counteracts the unsupported end of the corbel. The shorter stones are used first with the larger corbels on top to ensure proper cantilevering. Corbelling has been used effectively in India and other eastern countries for arches, vaulting and barrel-vaulted roofs.

Arch and vault-type roofing

There are many varieties of roof based on arch and vaulting techniques, which have been used for centuries in North Africa, the Middle East and Asia. Although such roofs are based on the use of locally available raw materials, the construction requires experience and skill, as is apparent from the photographs.

The pillars in a series of arches are usually square. Two rectangular blocks, with packing in between, form each course The cut stones are placed alternately one way and then the other. The external and internal radii of the stones forming the arch are parallel. They rest on the facing of the top course, while above the arches the wall descends in a point towards the pillar, to form the tympanum. Arches are almost always broken – in the past they were completed at the top by the vertical plane between two arch stones; today, masons seem to prefer using a keystone.

Floors Floors may be paved with stones, either laid on the joists themselves or straight on the ground. Sometimes in attics, low walls of stone can be seen, built with lime mortar on the floor to make small grain stores. There are some floors made with joists between which flat pieces of stone are placed tightly wedged on edge.

The use of stone tiles has considerably increased since the mid 1980s due to industrialization and the use of automatically controlled stone tile cutting and polishing lines.

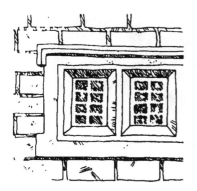

Cut-stone lintel

Important classes of end-use

Type of use	Equiv. sq. m (000)	Incidence (%)
Flooring	100 000	34.7
External facing	65 000	22.6
Funeral products	42 000	14.6
Stairs and halls	22 000	7.6
Structural works	22 000	7.6
Internal facings	15 000	5.2
Non-serialized products	12 000	4.2
Special works	10 000	3.5
Total	**288 000**	**100.0**

Above: *The lintel over this doorway in La Rioja, Argentina, is made from solid granite. Note the stones in the lower courses which are usually larger to increase stability.*

Right: *Roof-tops made from limestone building stones.*

Below: *Street pavement under repair using squared granite setts.*

[122]

Above: *Basalt stones, quarried locally, were used to build this interesting corbelled roof from Syria.*

Below: *The Inca base of a building in Cuzco, Peru, achieves remarkable fit without the use of cement or mortar.*

Cutting sandstone with a water-jet is semi-automatic and often speeds up the stone-cutting process.

8 Industrial production

It seems paradoxical that an industry which has been in existence for as long a stone has, should only recently, in the latter part of the twentieth century, begin to adapt itself to modern times. Even in technologically advanced countries, the use of traditional methods has persisted, some dating back to the day when a progressive mason became tired of his rough stone walls and squared his first stone.

The days are long past when an architect had to design stonework to conform to a limited range of shapes. Modern cutting techniques coupled with advanced fixing methods enable almost any design requirement to be realized. Wire-saw and band-saw cutting can provide any curved surface without the need for elaborate chiselling, and hollow diamond bits can cut columns which, in turn, may be speedily fluted by highly efficient shapers. High-strength adhesives, cements and fixing devices can hold stone in place where formerly sheer bulk was a deterrent. The provision of different colours and textures is limited only by the economics of transport.

The influence of recent technological innovations on architectural thinking is evident when one considers buildings such as the UNESCO headquarters in Paris, where modern fabrication methods allowed great flexibility and variety in the use of stone materials. The polychromatic application of granite to the façades in the renovated Bishopsgate and other city projects in London's banking district, and the numerous shopping malls with their 'urban furnishings' in stone to be found globally are further examples.

A major problem confronting the stone industry is manpower; in terms of mechanization it is far behind many other industries, with its smaller turnover and, at present, its more restricted usage. Few courses exist for training stoneworkers. In the past, apprenticeships were considered to be sufficient formal training in the stone industry. Unlike other fields, there were no technological changes in the raw material itself to require additional specialization. The irregularity in demand for stone during this century has left its mark on skilled craftsmen, many of whom have quit the industry, with no replacements coming forward. In some developing countries with little tradition in stone and an increasing demand for stonework, masons with experience in architectural work are altogether unavailable.

Recent mechanization has brought in its wake many problems. The stone operator of today has at his disposal elaborate machines including numerically controlled and fully computerized systems, capable of a high degree of precision to meet specifications. The variety of stone available has increased considerably as modern transport makes possible the delivery of any exotic variety, within reason. Delivery times have become more important than in the leisurely pace of yesterday.

All this coupled with increasing labour costs has resulted in an attempt by the industry to plan its future with more emphasis on sophisticated machinery. Automation has become the main target to make dimension stone competitive with other construction materials.

It is incongruous that among the materials which have replaced stone, there exists a whole range which are based on rock products. And yet it is less expensive to quarry rock, break it up and put it together again than it is to obtain the end product by direct fabrication. The stone industry has perfected qualities of reconstructed stone, artificial stone and even reconstituted stone. All of these are based on rock products – cement and stone. To a certain extent cement has been replaced with plastic binders in marble construction, resulting in a very high ornamental effect. All this helps the natural stone industry by providing an outlet for its secondary materials, thus reducing the cost of the natural product.

Comprehensive data on rock and stone properties, gathered in the engineering geology discipline, is available in journals to the stone industries and has resulted in the design of equipment and working technologies, especially quarry technology, as described in Chapters 4 and 5.

Quarrying practices

Basic practices have been discussed in Chapters 4 and 5. Industrialization in quarries, as compared to processing is, in a sense, an extension of practices used in smaller workings. The compressor is a basic piece of equipment in most stone quarries – in the larger quarries, compressors have a larger capacity; compressed air is piped throughout the quarry, powered by an electric power installation, either country grid or local generator. Tools are larger and heavier – often fully hydraulic – and where possible several are operated from centrally located commands, especially drills. Hydraulic wedges, hammers and chisels, diamond wire-saws, large chain-cutters and self-propelled jumbos are standard equipment in many of the large quarries.

Drilling methods

Compressed air drilling has been, and still is, the most versatile means by which to extract stone materials, even though it may not be the most efficient industrial process. Multiple drills mounted on quarry bars, wagon drill types, self-propelling drilling machines, jumbos capable of drilling holes vertically and horizontally at the same time by using jack-hammers in the required positions; all of these are basic equipment. Pneumatic drilling practices are summarized in Chapter 4 as these smaller systems apply to small-scale as well as industrial extraction. Drilling is so well entrenched that it is considered a traditional system. The Finnish system is based on hydraulic equipment and highly controlled blasting.

Some operators use the Swedish system where holes are drilled contiguously in a row, thus

Circular saw

obviating the use of wedges and the risk of an uneven break. A bar is always inserted in an adjacent hole during drilling to ensure straight holes. The Swedish system involves a variation of the drill-and-broach method used in driving horizontal channels, whereby closely spaced rather than touching parallel holes are drilled and the space between the holes is broken out with a broaching tool. Less common nowadays is the use of a radiolax machine – a pulsating piston drill operating a bit that is mounted on a swivel that can be swung up and down in a vertical plane.

Channelling is sometimes used instead of drilling. Power-driven channelling machines move over a rail placed on a quarry bed. The channel is made by the chopping action of a set of cutters similar to that of a reciprocating drill. Quarry bars, where air drills mounted on saddles move along a steel tube supported by logs or skids, have to a certain extent replaced channelling machines, especially in smaller quarries. The Knox method of quarrying is little used but may prove useful; light charges of black powder are placed in a line of reamed holes. An air space is left between the charges and the tamping so as to spread the force of the explosion over a large area. The reaming that has been executed in the direction of the required splitting promotes a straight break.

Wire-sawing It is likely that more marble is produced by wire-sawing than by any other process. In this method an abrasive – sand or special granules – is carried by a helicoidal steel cable. The wire is moved by a motorized drum and cuts through various stone surfaces with the help of support pulleys; a tensioning device ensures steady progress. Previous drilling is required to insert the pulley support brackets. The advance of the wire within the rock mass is ensured by a helical screw which drives the main pulleys, controlling the entrance and exit of the cable. Long, short and medium circuits are used, depending on performance requirements. The wire may consist of single, double or triple threads. At one time this method was commonly used at the processing plant for many materials including granite, quartzite and sandstone, for shaping and cutting thick slabs. In quarries, it is now an accepted method for cutting rocks containing little or no quartz.

This type of wire-saw is being superseded by the use of diamond beads mounted on a wire for the extraction of marble at the quarry although the method is still in the early stages for the processing of granite. The equipment is powered by 30 to 60 hp when an electric motor is used. This is preferable to a diesel engine, which requires some 80 hp. The length of the wire varies from 20 to 80 metres and requires 1000 litres of water/hour at normal water pressure. Cutting rates of 5 to 10 square metres/hour have been obtained; the higher figure is for travertine. The equipment can be operated from a remote control console. Advantages of using a diamond wire-saw include:

o The reduction of wastage due to straight and smooth cuts.
o No squaring operations required for fitment into the gang-saw.
o Comparatively high-speed extraction.

Extraction by chain-saw Chain-saws or chain-cutters were at one time standard mechanical equipment for coal-cutting in collieries. From the 1960s they were adapted for stone-cutting in quarries and after overcoming frequent problems caused by machine breakdown and the need for sharpening and servicing, they are now widely used, especially for the opening of underground faces.

The cutter is powered by an electro-hydraulic motor with speeds of up to 8 linear metres/minute. In the larger models, the arms usually make a vertical cut of 2 metres and horizontal cuts of 3 to 3.5 metres. The cutting blades are made of tungsten carbide and can be flipped over to obviate the need for sharpening during their operative life. The cutters work best in homogenous rocks like travertine, peperino, tuffaceous rocks and botticino-type limestones. The presence of quartz or other hard minerals is an obstacle. Other restrictions include the limited depth of cut, the weight of the cutters (3.5 to 4.5 tonnes) requiring special lifting equipment, and the time taken to reach the full vertical cutting position which is more than double the rated cutting speed of 8 to 14 square metres/hour, depending on the material. The combined use of chain- and diamond-cutters is the most efficient application.

Flame-jet cutting

This method has been used since the early 1960s in igneous rock quarries and can only be employed with polymineralic rocks. A concentrated high temperature flame (2000°C) with a high heat transfer capacity exceeding 40 000 kcal/hour is confined to a narrow strip of rock giving a cut about 10 cm wide. The more varied the expansion coefficients of the different minerals, the more efficient the method, especially in acidic quartz-containing rocks. Fusion may cause re-cementation on cooling and is to be avoided.

The use of kerosene and oxygen has been replaced by the less explosive compressed air/diesel fuel combination. Consumption includes compressed air at 7 atmospheres with 4 to 8 cubic inches/minute, 80 litres of diesel fuel/hour. The average nozzle life is 700 to 1000 hours and the cutting rate is 1.5 to 2.5 square metres/hour. Normal cuts reach a depth of about 5 metres, although cutting depths of up to 20 metres in quartz-rich coarse-grained granite have been reported.

Flame-cutting can also be economical for boring wide-diameter holes in sandstone, quartzite rocks and dolomites. A major drawback is its high noise level, which may exceed 120 decibels, and which makes the method impossible to use in built-up areas. A similar method is used for scabbling blocks prior to placement in frame-saws, using a standard blow-pipe nozzle with liquid gas (a methane/propane mixture) with the addition of oxygen as a combustion aid.

Water-jet cutting The experimental cutting of stone by water-jet has been reported since the early 1970s. It took more than a decade to overcome the difficulties involved, especially in the cutting of harder rocks where the method would have been most useful.

Water-jet cutting is industrially applied to undercutting solidly bedded (10 to 12 metres thick) sandstone in North Vosges and Rothbach (Bas-Rhin). The equipment used there includes a 5000-litre mobile water tank with a 650-bar pump. The jet heads are operated with a 10-metre-long horizontal support, which can be moved up and down to a height of 10 metres, making a horizontal cut 8 to 10 cm deep. The water stream reaches a speed of 1300 km/hr under 650 bars. The orifice moves slowly to and fro along the 10 metres, cutting to a depth of about 50 mm per return movement. The cut is made with an oscillatory movement (20 cm amplitude) and a 3 square metre/hr output is achieved with a water consumption of 60 to 80 litres/minute. Similar results have been achieved with granites and quartz-bearing porphyries, but with higher speeds (2500 to 3100 km/hr) and at pressures of about 2500 atmospheres with a cutting depth of 20 to 25 mm at each pass with 2 square metres/hr cutting output. The cutting speed can be improved by using a mixture of water and abrasives.

Processing In line with the desire to increase production per machine unit, harder cutting edges and media have been investigated. The last few decades have seen a rapid development in the use of diamond and tunsten carbide tools, whether in connection with cutting blades, cutting discs, or as grinding media.

Diamonds have a definite role in the industry, the problem being where to use them and to what extent. Marketing of diamonds and quantitative data on their use has not caught up with the fast technical developments of tools. Whereas the diamond tool industry is putting much thought and research into the mechanical improvement of their products, operational data is difficult to standardize: agreement on cutting speeds is an example. The standardization of tools by the various manufacturers would facilitate comparison between one product and another, and performance could be compared with the prices of diamond tools. This is especially important when considering that a 3500 mm disc costs tens of thousands of dollars.

Apart from their use in extraction, diamonds could be useful in primary cutting. In secondary cutting, discs already seem to have fulfilled existing needs to a large extent, whether in sedimentary or igneous stone.

At present, many stone firms have to carry out their own tool development research, aided by tool producers rather than machine manufacturers. However, the co-operation between the machine, tool and stone industries is improving. At one time a major point of contention was who would pay for research and development. Any increase in the use of stone will provide encouragement to both tool and equipment manufacturers, who have to deal with a very specialized market of a limited nature, complicated by variations in material from quarry to quarry. Fundamental properties of stone are being studied quantitatively and are now available as standards, *agréments* (EU) and codes of practice.

Primary cutting For sedimentary rock, the use of frame-saws and gang-saws is still considered the most economical means of subdividing blocks into slabs and other rectangular sizes. The main development in cutting has been in the application of improved diamond blades, discs and wires. Until recently, blades in gang-saws were fed with abrasives in suspension, usually water, which acts as a coolant. The introduction of high-cost diamonds added a factor in the running costs of gang-saws which may have been neglected with conventional cutting materials: that is, the preservation of their waste, which was of comparatively minor importance, even if the abrasives were expensive, high-energy materials.

Data to compare costs with abrasive-fed blades is variable, but there seems to be agreement that diamond blades are often uneconomical for slabs less than 2 cm thick and for certain stone stypes. Again, little data is available on the mounting of cutting edges, the size and quantity of diamonds, the lubricants and coolants which should be used and their temperatures. The tensioning of diamond discs has been improved, although there still exists the problem of conducting enough coolant to the diamond cutting edges. The use of a hollow blade assembly consisting of lower blades welded into a third upper blade, looking in section like a tuning fork with diamond segments brazed in between the prong terminals, has been proposed among other ideas as a solution, but welding distortions have prevented the development of this device on a commercial scale. Bonding by laser welding has opened up new possibilities – this technique is now used for dry-cutting discs and chain-saws used in secondary sawing.

For harder stone, particularly of igneous origin, wire-saw machines are often used, the advantages being the elimination of staining on light-coloured materials by using diamonds or iron-free abrasives, the ease with which angles can be cut around the axis of wire and superior cutting speeds. The reluctance to use diamond blades in granite is due to the presence of hard and soft grains, which result in a jerky action on the cutting edge, thus reducing its life. However, the successful use of diamond blades in igneous rock with minerals of equal hardness has been achieved.

Much thought is continuously being given to the development of a vertical gang-saw. The advantages over the horizontal system are as follows:

o Blocks of considerable length can be cut. Their length is limited only by the distance the carriage carrying the block can travel to feed the saw.
o Finished slabs can be removed without stopping the saw.
o Cleaner cuts can be obtained by virtue of the rigid shorter blade.
o A continuity in ornamentation, for example of horizontal elements, can be realized, adding to the possibility of architectural uses for stone.

Several other varieties of gang-saw have appeared on the market and the tendency is towards heavy construction to minimize vibration and thus save time in grinding by starting off with smoothly sawn surfaces. Developments include gang-saws without a central beam so that blocks can be handled from the front to the rear of the machine.

The vertical diamond band-saw is useful when cutting single slabs from a block when an urgent order has to be filled. It is usually not used in regular production except, perhaps, on very expensive material, as the cut is only $\frac{1}{8}$-inch thick and saves material. A single diamond blade mounted in a special frame is being increasingly used industrially; single blades (monolamas) fed by sand and other abrasives have been known and used for a long time, sometimes with the sole purpose of trimming blocks to prepare them for the gang-saw. For softer stones, tungsten carbide chain-saws are used.

Gang-saw blades for loose abrasives are sometimes perforated by holes, staggered both horizontally and vertically. They may contain diagonal channels and serrated teeth, both rectangular and triangular, or with a slight corrugation, which retain abrasives and prevent blades from jamming.

Another development was the mounting of large diamond discs in a frame. This is a flexible way to cut blocks without tedious size-matching, in order to fill the capacity of the gang-saw.

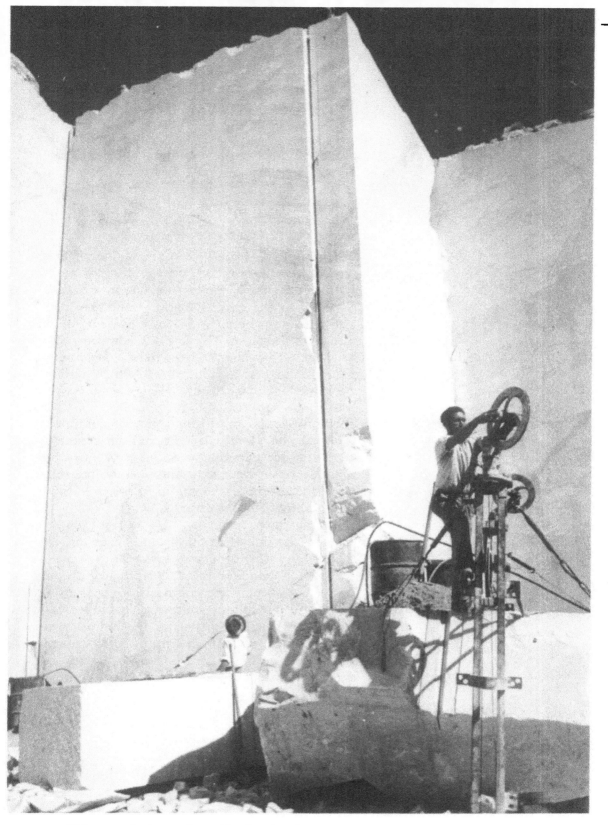

This type of wire-sawing with abrasive slurry is fast being replaced by diamond wire-saws.

Single or multi-disc frames are often used as auxiliaries to gang-saws where small or irregular blocks are quarried and where high scabbling costs turn a high percentage into waste; in one of these frames, a disc of 3 m diameter is capable of cutting slabs of 0.80×3.6 metres to any required thickness and can be pre-set for automatic operation. Production achieved for a 2 m disc is 7 to 8 square metres/hr on a squared $2 \times 2 \times 3$ metre marble block.

The nature of the lubricant may be of great importance in primary cutting: it must be the right temperature; its wetting factor should be matched to the cutting medium and the material to be cut; the right pressure must be determined, as a tendency to give a maximum supply of water to the diamond segments may cause a pressure pocket to be formed, which reduces cutting efficiency.

Secondary sawing

Diamond-saws have now almost completely replaced the slower silicon-carbide carborundum-saws for trimming stone after primary cutting. However, because diamond-saws are more expensive, some carborundum-saws are still in use, mainly with power tools.

Again, data varies on the correct cutting speeds, the ideal amounts and sizes of diamonds and the proper matrix. They have to be carefully matched to the petrological properties of the stone. The emphasis is generally on automatic machines. Multi-disc machines have caused production problems in the past and discs should preferably be mounted to separate spindles. Although much improvement has been achieved, the construction of common drives requires a high degree of precision, coupled with the uniformity of the diamond segments to ensure even wear.

Cut-off sawing machines

The machines generally use rotary discs and are operated in two ways:

○ By moving the tools on rails or bridges and keeping the material stationary except for the rise and fall of the workpiece.

○ By moving the stone or rather the supporting table on rails and keeping the cutting unit stationary, except for the rise and fall.

However, for multiple production work, rise and fall are not essential. The difference in cutting action between the two possibilities is small, although the choice of system may be relevant for production purposes. Both systems require pressure control during feeding which may be power-driven or manual. The size and speed of the motor determines the diameter of the cutting disc; a variable trapezoidal-belt drive and the provision of interchangeable pulleys allows the variation of disc sizes. The disc spindle is usually at right angles to the long axis of the supporting table – swivel heads are adjustable in several directions.

The cutting depth can be fixed by raising or lowering the cutting unit. The thickness can also be pre-adjusted to ensure even widths during the cutting. Many types of sawing machine can be used for milling and moulding operations. The use of 'tagliablocchi' type sawing machines with horizontally and vertically mounted discs is now standard. They are often used for primary cutting, especially for small and irregularly shaped dimension stone blocks.

Surface processes

The term 'polishing machine' is used in the text although a large part of the work is roughing, gritting, grinding or honing and only the last stage is actual polishing.

Smoothing or gritting of the stone surface after sawing is achieved by stepping down the grade of the abrasive in three to six stages, depending on the degree of finish required. Semi-automatic and automatic machines with forced feeds are replacing hand-operated surfacers, which are favoured where a product does not warrant larger machines. Generally, rotating discs are employed to carry the abrasive surface.

Diamond set or impregnated surfacing rollers are used in a few plants. Polishing is used to bring out the full character of the stone. The durability of a polished surface is commensurate with the care given during the processing. Polished surfaces for exterior work are only recommended where no doubt exists as to the reasonable duration of the finish without undue maintenance.

Surface treatment after polishing is often applied to improve the gloss, and to protect finished slabs during transport and handling. The desirability of a waterproof backing or protective surface treatment, for example silicon, is controversial.

Gritting
About a century ago, a polishing machine with mechanical arms was the first step in the mechanization of gritting operations after the slabs leave the saw. Until then smoothing was mainly a hand operation, and still is in remote areas where no power is available. The 'Jenny Lind' radial polisher, named after the famous Swedish singer, is still in use today and is a prototype of the arm polisher. A rotary polishing disc mounted on a universal head can be swung by an operator over the stone table, to and fro and back and forth. The use of two arms pivoting on joints with bearings enables a wide sweep. The arms are mounted on brackets attached to a wall or to a pillar. Cylindrical pillars allow the arms to revolve through a full circle continuously, reducing unproductive mounting time. Water is fed through the polishing spindle which slides up and down in a metal sleeve built into the head at the arm. Power is transmitted by belt or by direct drive from a single or a multi-feed motor. Pressure is transmitted through a handle with pivot holes at several positions to allow maximum operating pressure whatever the slab thickness, and adjustable to the most comfortable height for the operator.

A self-retaining catch allows the polishing handle and weight of head to rest whilst discs are being changed. The universal joint between the rubbing head and the spindle allows the head to take up any irregularities in the surface.

Grinding blocks are fixed on the disc by shel- ' lac, polyester adhesive or other binders, or else are fitted into a special fixing device. First a coarse cutting-stone is used to smooth off the raw edges left by sawing. As the stone becomes smoother, a finer grade of abrasive stone is used, until scoring marks are no longer visible. Occasionally, with materials of a coarser grain, the full polishing sequence as described above may not be followed, and once the stone has been made absolutely smooth the actual gloss may be achieved by a polishing agent, rather than a polishing process.

The arm-polisher has a very low output, and requires a man in full-time attendance to move the polishing head over the stone. Automatic polishing machines work on the same principle of interchangeable facings to a rotary polishing head, but are designed to track automatically over pre-set distances, from front to back and from side to side, across the entire surface of the stone within the compass of the bed of the machine. They have a very much higher output than a manually operated machine, produce an equally good finish and a large number of machine heads can be efficiently controlled by one man.

However, for certain tasks arm-polishing machines are important especially if forced pressure devices are incorporated. Models for granite are usually more powerful and can be used for planetary heads.

Placing more than one thickness of slab at a time on a polisher, whether hand-operated or automatic, should be discouraged as this causes the grinding stones to be chipped as well as damaging bearings and alignments although, in polishing machines with sensors, the tool is lifted before it reaches an edge and various thicknesses can be surfaced at the same time.

Finishes   There are no accepted international definitions of the various finishes of fabricated stone. The following terms for the usual marble finishes are applied by the Marble Institute of America, mainly for industrially processed stone.

Natural finish:	A finish produced by sawing.
Sand finish:	A finish produced by sand-rubbing.
Sand-blown finish:	A finish produced by sand-blasting.
Grit finish:	A smooth finish.
Hone finish:	A velvety smooth finish with no gloss.
Polish:	A mirror-like glossy finish which brings out the full colour and character of the marble. When used, periodic maintenance is required to retain this finish.

Polishing Problems connected with roughing or grinding have been overcome to a certain extent with the use of automatic machines. However, to obtain a high polish by present methods requires no less attention than that needed with hand-operated surfacing machines. Whereas some advances have been achieved in other stages of the fabricating process, polishing has remained very much unchanged. Practices vary from locality to locality, and from one variety of stone to another, but the general approach is very similar.

Polishing is the most important finishing stage as it brings out the full character of the marble and granite and the maintenance and durability of the surface depends much on the quality of the polishing. Little is known about the mechanics of polishing. Since materials used in polishing are very fine and often softer than the materials being polished, it is likely that they do not act as mere abrasives. Polishing was originally thought to be a matter of fine grinding, in which ridges and depressions left by a coarse abrasive were reduced to suboptical dimensions by the use of finely divided polishing agents.

It has been postulated that the friction during polishing causes an actual fusion of minute projections resulting in a flow, that is, an extremely thin liquid layer caused by local heating

Stone polishing machine used on marble slabs quarried and cut locally in Haiti

or chemical action with the production of an amorphous surface layer. This so-called 'Beilby layer' may cover and hide scratches due to abrasion during the grinding stages. Electro-diffraction methods have confirmed that in calcite, the Beilby layer re-crystallizes immediately after formation on the cleavage planes, but remains amorphous elsewhere.

It has been suggested that the relative melting temperatures of the polishing agents and the material to be polished is relevant, rather than the relative hardness. For example, calcite was rapidly polished by polishing powders which had a melting point higher than its own. Calcite (melting point 1333°C) showed rapid polish and surface flow with zinc oxide (melting point 1800°C) but little with cuprous oxide (melting point 1235°C) under similar pressures.

Different stones require different polishing media and it pays to take great care in finding the right compound, as much time can be saved in obtaining an optimum finish on the production line. With improved micro-size control, fine diamond powders are increasingly being used for polishing and honing.

At times, the direction of polishing and the right pressure is of relevance, especially in veined marbles where undue pressure tends to open somewhat softer veins. Factors such as the speed of the buff, degree of moisture and quantity of powder used may affect the quality of polishing. The most commonly used polishing agents are oxalic acid, tin oxide and aluminium oxide. Oxalic acid is popular because it produces a high finish with little labour, but the polish is also less durable. Tin oxide or finely-ground metallic lead mixed with putty powder (an undefined product) produces a high and durable gloss.

Fixing As large quantities of stone are used in cladding, much attention is being paid to this area in order to compete with other materials. Stone facing should be as thin as feasible and fixed in the largest possible units. The cladding should be safely attached to the constructional framework, particularly in tall buildings. Prefabricated panels lend themselves well to fast and effective fixing, and a whole range is becoming available. There has been considerable comeback in the use of granite for facing, as this is a most durable material and, unlike stones of sedimentary origin, it can be used out of doors for highly polished surfaces.

The common factor of all the fixing devices and methods available is that one end of the device has to be anchored or keyed into the slab. This requires a minimum thickness of marble to hold the device without breaking. Whether the anchoring is done on the site or whether the stone is backed on to prefabricated units, the problems remain very similar.

Integral prefabricated units are easier to control under plant production conditions. Whatever the material is to which slabs are bonded, the bond has to be safe and permanent. An ideal prefabricated stone panel should consist of three main elements. The exterior part should be decorative, wear- and weather-resistant. The middle portion should make up for the relatively low thermal insulation property of stone. The interior part is mainly decorative, if used as a wall. In curtain walling the interior part does not require any special finish and can be left as sawn.

Automation The automation of production should ideally be coupled with the standardization of stone units. Modular units make stone more competitive with other building materials. Another advantage of the modular unit is that it can be based on standard frame-saw dimensions, as the maximum use of individual stone blocks is an important factor in dimension stone economics. An expensive item in stone fabrication is the setting of tools on to machines; wear through abrasion requires frequent adjustments which are costly and also delay production.

The ideal flow line for dimension stone production would be the following:

- Quarry with a bed thickness of 1.2 metres.
- Frame-saws with practically no vibration, capable of cutting slab thicknesses of 8 mm.
- A battery of roughing and smoothing rollers to prepare the stone for final polishing.
- An automatic stopping or filling device to seal holes or seams.
- A high polish set-up, which would work at the same rate as the rollers, thus obviating an existing bottle-neck which is overcome by varying the proportion of grinding to smoothing rates of polishing heads.
- An edge smoother and polisher synchronized to the above.

A summary is given here of the past and recent practices more frequently encountered in the final finishing stage:

- The surface before final polishing has to be flat and free of scratches and pittings.
- Oxalic acid takes first place in polishing marble; tin oxide in granite. Salts of sorrel may be added to the oxalic acid. A small amount of sulphuric acid, and in some cases tin oxide, is also added. Where the chemical composition of local water resources requires, tin oxide may have to be used in the polishing of marble. Improved polish is claimed when oxalic acid crystals are ground to a very fine mesh.
- The polishing medium is usually applied by a felt buff of medium hardness; sisal or jute in various forms may be used. Cast metal discs containing a polishing medium are available, too.
- The application of submesh diamond powders is increasingly practised on a commercial scale in the stone trade.

Sand-blasting This finishing method is applied in most countries as a surface treatment for granites and as an auxiliary method for lettering and ornamentation in marble. In Sweden, especially, it is also applied as a main surface finish for marbles. In most cases, this finish is obtained by using abrasives other than the original sand which gave its name to the process.

Thermal finishing   There is no established standard practice for the use of flame-jets in the finish of granite surfaces. Experiences vary, but many consider a thermal finish to be a more natural surface than a highly polished one. Staining experiments have been conducted in conjunction with thermal finishing and some very effective colourings have been obtained. Thermal finishing, at one time done by hand, is now mechanized and electronically controlled.

Water-jet   Experimental cutting of stone with a high pressure water-jet has been reported since the early 1970s and is now used in a few quarries, notably for sandstone, for example, at Rothbach in the Northern Vosges region of France. The method is increasingly used in processing plants for all types of stone for intaglio, precision cut-outs and shaping slabs.

Mechanical stopping   A stumbling block in the automation of high-grade finishing processes is the need for stopping or filling of holes, especially in coloured marbles or travertines. This is usually done before the final buffing stage, and mostly by hand. Mechanized stopping is available and incorporated to a certain extent into some fully automated processes. The most popular material for stopping is white cement with appropriate colourings, although polyesters are being used to great advantage both for stopping and repair work on the finished table.

Waxing and packing   In some cases, surface treatments are adopted to improve the gloss, but are mostly used in the protection of the finished marble slabs during their transportation to, and handling on, the fixing site. The traditional beeswax and turpentine mixtures may be used, although in industrial practice these have been replaced by a whole range of proprietary products. Some firms do not consider slabs for cladding finished without giving them a waterproof backing. In curtain walling, outside waterproofing of the interior wall is practised.

It is now an accepted practice to use nylon backing such as netting or gauze on problematic stone slabs, particularly with thinly cut, strongly coloured varieties or stylolites.

The expected development of the stone industry will result in a large expansion of new plants and equipment in relation to existing establishments. The running-in of a plant is just as important as the running of a plant. For this reason any new plant, or plant installing new capital equipment, should have an experienced processing expert to guide the plant or machinery operators. Inexperienced operators should never be trained on new equipment as this increases maintenance costs. Far better to leave the equipment idle and maintain it systematically until it is needed for production. There is a tendency to acquire the latest models in plant equipment without taking into account the need for highly skilled supporting services, including sound foundations, proper levelling, controlled abrasives and frequent maintenance. The overloading of electric motors, together with general misuse of equipment, is frequently observed in new plants. Also, equipment, however up-to-date, does not absolve one from having to apply meticulous quality control to meet modern requirements.

Above: *Sawing limestone into building blocks; industrial blocks have to conform to exact measurements.*

Below: *Preparing marble blocks from Carrara for shipment.*

Above: *At this marble quarry a diamond cut-off sawing machine in the background operates from an electric generator, while in the foreground drilling and wedging operations take place.*

Below: *Gang-saw blades making slabs out of a marble block in the Philippines.*

[148]

Above: *Diamond-sawing machine at stone pilot project in Port au Prince, Haiti.*

Below: *Diamond band-saw slicing a marble block.*

9 Stone development

The basic requisites for the establishment of a stone industry are the existence of accessible deposits with proven reserves and a market for the stone produced. These factors are just the starting point; for a stone industry to flourish and prosper, other areas are important, too, including marketing, national planning, forecasting, costs, and a knowledge of extraction and processing techniques. These points will be discussed in this final chapter.

Comparative costs

Many developing countries have a special interest in the more widespread use of stone resources, particularly those dependent on fuel and cement imports. Even before the huge increases in energy costs, building stone was cheaper than concrete blocks of a comparable size in a number of countries. Building with concrete means using materials with a high consumption of energy in their manufacture, such as iron, steel and cement. Energy accounts for as much as 60 per cent of costs in the production of cement.

Similar high energy costs are found in brick production. One kilogram of wood is needed to produce a single kiln-dried brick; a small cottage industry producing 10 000 bricks would require 10 tonnes of wood. Conservation is a key factor, too, as brickmaking contributes significantly to deforestation and erosion. Forests could be saved if stone replaced wood-fired bricks as a construction material.

Apart from low energy requirements, building stone has a number of advantages over manufactured construction materials: low initial investment, labour-intensive techniques that can be freely adjusted to meet needs and the flexibility to allow small-scale production units to be set up. These factors add up to a low-cost building material which could rapidly make a significant contribution to building construction.

Planning In recent years a new concept has been introduced: integrated stone development planning (ISDP). This covers the development of stone from planning to marketing. The pattern is the same for developing and developed countries alike, but it is easier to implement in the former, which have informal construction traditions; they are relatively free from legislative and planning restrictions, and are disposed to consider practical rather than traditional standard specifications. Integrated stone development planning progresses along the following lines:

○ Make an inventory of existing and potential stone resources.
○ Include the inventory in national planning to ensure conservation and the proper use of resources.
○ Consider environmental and energy-saving factors.
○ Introduce appropriate technologies and improve current practices.
○ Establish central testing and documentation facilities, practical Standards and Codes of Practice.
○ Train personnel, on a national basis.
○ Introduce stone products into local low-cost housing schemes and road-paving projects.
○ Upgrade stone for industrial and export purposes.
○ Initiate new products based on stone as a raw material; for materials using fillers and manufacturing rockwool, for instance.
○ Intensify the use of lime mortars and similar materials.
○ Establish small-scale production units.

Institutional interest in integrated stone development planning is essential, if only because stone deposits are national assets, non-renewable and with a strategic value. No government can afford to ignore the conservation of such resources.

Regional and international marketing are also important considerations in stone development planning. The standardization of terminology, statistics, definitions, specifications and testing procedures would facilitate international co-operation and allow marketing studies and forecasts to be made on an international basis. Developing countries on the threshold of establishing a stone industry could thus benefit from the experience gained by countries with long-established stone industries.

The ideal plan for dimension stone development is to establish integrated stone development planning within an infrastructure supplied by the national or local government. This creates the sort of climate that attracts investors. It is relatively easy to open a quarry – hence the ubiquitous 'blot on the landscape' – but more difficult to ensure its proper development in economic and environmental terms.

Investors usually arrive on the scene when the first positive results of the inventory work become known. Although the best time for investors to move in would be when the inventory has been completed, in practice claims would already have been staked by that time.

Investing in equipment When a stone deposit is reasonably accessible to the potential operator, just a few dollars' worth of tools can produce saleable *objets d'art* or more useful items such as mortars and pestles. Where time is not an important factor, many a house has been built from a stone deposit *in situ*. When the deposit is a slabby one, and the stone possesses the necessary physical properties, a low-cost housing scheme could be established. Drilling rods, hand levers, hammers and chisels, a cutting guillotine, possibly jacks, and simple mechanical haulage and lifting devices would be sufficient equipment to build sound and solid, if simple, stone dwellings. For more sophisticated purposes, a sawing machine would be required to produce lintels, tiles, sills, wall and roofing components.

Marble and dimension block production

requires compressors, as well as the accessories already listed. In addition, lifting, haulage, transport and delivery require heavy equipment, which steeply increases investment costs.

An important factor is the nature of the deposit: ease of extraction, workability of the stone and so on. Generally speaking, it should be possible to establish an ornamental rock quarry producing 1000 cubic-metre blocks – a turnover of hundreds of thousands of dollars – with an equipment investment running into tens of thousands of dollars. The working capital required depends on more complex factors.

Experience has shown that the industrial development of stone in developing countries should take place gradually. Setting up large-scale extraction and processing operations is encouraged only in very exceptional circumstances, with well-proven suitable reserves and an assured demand.

Commercial factors The size of industrial blocks ranges between two and three metres in length and one to two metres in width. A thickness of half a metre to one metre is the usual minimum requirement: maximum size depends on the capacity of the gang-saw. Blocks smaller than a certain size will command a lower unit price. Measurements aside, blocks for industrial use must be free from hollows or protrusions.

For many decades, the standard thickness of sawn stone slabs was 2 cm, but this is now being reduced. Sawing granite into slabs 1 cm thick is a common practice, even for export, where improved packing techniques have reduced the dangers of breakage during transport. When stone is to be shipped, primary sawing into slab form may bring to light natural defects not obvious in block form. Any additional handling and packaging costs will be compensated for by avoiding time-consuming claims and litigation which may result from the delivery of defective material.

Quality control has assumed greater importance because of:

o the demand for standard sizes, often with rigidly laid-down tolerances, essential in pre-fabrication and modular building techniques.

o consumers who are reluctant to abandon established producers with good reputations for new suppliers unless prices are attractive enough to cover any risks.

Pricing The pricing of stone depends upon a number of factors, including grading, demand and availability. The Marble Institute of America has classified stone, and marble in particular, into four groups. Group A includes marble and stone with a uniform appearance and favourable working qualities. Marble and stone in Group B are similar in character to those in Group A, but have less favourable working qualities. Group C contains marble and stone with inconsistent working qualities; geological flaws, voids, and veins are common in these types. Group D contains materials similar to those in Group C, but with a larger proportion of natural faults and greater variation in workability.

Prices per square metre for polished sawn slabs of, for example, travertine have changed little recently. Group A and Group B varieties are US$12 to 23, US$30 to 42 and US$27 to 35 respectively. If a greater thickness is required the cost will increase by 40 per cent per centimetre thickness.

At one time, high-grade limestone for industrial use was a by-product of dimension stone workings. Nowadays it is often a main production line, exceeding low-grade dimension stone in unit price; indeed, some varieties even have export potential.

High limestone prices have, in fact, accelerated the trend towards granite, which is gaining in demand on the world market and making a notable impact on the stone trade. No definitive grading system has been established, and factors such as transport, handling, breakage, packing and so on affect prices, too.

Transport Standardizing sizes has made stone consignments more suitable for modern forms of transport, resulting in cost savings. Door-to-door truck deliveries from plant to distributor or consumer are available throughout continental Europe. Cabotage-type haulage is widespread and comparatively small boats deliver shipments to importers at the nearest inland port, although the use of containers is increasing.

The overall cost of transport must be measured in relation to the price of the stone. For rare marble and other ornamental varieties which command premium prices, transport and handling costs are less important.

It is interesting that the constant expansion of scheduled transport services will increase the availability of more exotic varieties of ornamental stone, often found only in remote areas which are difficult to reach.

Promotion The aims of promotion are twofold: to market the product and to attract investors. The first step is to gather relevant data on raw material reserves, production statistics, size of industries and economic indicators, for instance. This data is then used to locate trade and distribution channels and to test consumer preferences.

A useful tool for promotion is a catalogue, presenting an inventory of the stone resources available and explaining their potential. The author has been involved in the production of several publications of this type, following stone development projects in the Philippines, Haiti, Nepal and Brazil. *Marble in the Philippines* (Shadmon, 1969) increased marble exports by 500 per cent in about one year, and subsequent exports have included a lucrative contract to supply marble for a 16-storey building in the USA. *Stone in Nepal* (Shadmon, 1979) soon generated the first exports and investment interest, while *Stone in Haiti* (Shadmon, 1980) prompted many export enquiries and resulted in private investments within a year. *Stone in Brazil* was published in 1988 (Shadmon, 1988).

When ornamental features are important to a potential client, photographs should be made available for them to make an initial selection.

Storage (above) *and transport* (below) *are factors which must be considered when setting up a stone industry.* **Opposite:** *Great Zimbabwe* (see page 4)

Their final choice should be based on actual samples showing patterns, colour ranges and various dressings. Standard sizes and an impeccable finish are essential. When unfinished cut stone is being marketed, the purchaser might prefer to arrange for polishing to take place when the stone reaches its destination.

Regional co-operation in marketing could reduce the dangers of short supply which, if it occurs, may allow substitutes to displace stone and set back demand. International development agencies could assist by turning their attention to the promotion of trade in stone products, especially on a regional or inter-regional basis.

Statistics for the future
As in other fields of resource development, the collection of statistical data on stone is a basic requirement in furthering the progress of the industry. This information will help national governments, international agencies, equipment manufacturers and others to discern trading patterns, determine conservation policies and improve equipment. In turn, the quarry or workshop owner will obtain more suitable plant, a keener appreciation of marketing possibilities and information of an economic nature from which to plan production.

Only a few stone-producing countries maintain quantitative records and even these are seldom consistent. Different ministries within the same country use categories to accommodate their own particular sectoral responsibilities, thus data may not be compatible (or comparable). Problems also occur when data on stone and marble production is grouped with data on other mineral commodities.

On a quantitative basis, calculating the tonnage of certain stone categories is complicated, since worked marble and granite slabs, for instance, are usually described in square feet or square metres, and converting to weight may cause considerable errors. The large variety of stone by-products, the lack of common standards of measurement and the existence of grading systems which more often than not reflect personal tastes also complicate the gathering, comparison and interpretation of international statistics.

Estimates of the value of production are further complicated by factors such as fluctuating exchange rates, currency devaluations and different financial periods.

As stone statistics, and especially production figures, are generally incomplete at the present time, the best indication of trade flow patterns can be obtained from import and export data, although these are also somewhat distorted by local consumption, fluctuating demand, stockpiling, re-exports and other factors.

An agreed statistical system of reporting on stone production must be a top priority for the future.

Forecasts The scarcity and lack of uniformity of available data relating to supply and demand in the stone industry is a great handicap in the compilation of market surveys and forecasts. At the national level, centralizing the information and services needed to promote the development of stone resources is a prerequisite to steady progress.

A major objective should be the promotion of a country's stone resources as an ordinary construction material rather than as an unusual or exotic one. This requires a demonstration of stone's competitiveness, in terms of building costs per unit area, by comparing stone with other construction materials, taking into account its lower maintenance costs and other benefits.

Data fed into a centralized unit should come from all areas: quarrying, mining, engineering, geology, building technology, architecture, the fine arts, archaeology and other related disciplines. With access to information from these sources accurate forecasts can be made to assist with production planning.

Foundations of success To a fresh generation of architects and interior designers, stone has become a 'new' material to work with. Besides its traditional uses in facing, cladding, flooring, columns, mantelpieces, vases and ashtrays, stone is also being used in prefabricated units, kitchen counters, and so on.

Yet the share of developing countries in the stone trade is negligible compared to its potential. Most countries have stone and rock resources which can be developed and exported, although often these resources have not been identified with the result that governments are not aware of the potential. Indeed, in some countries stone and rock are the major, if not the only economic mineral commodities which can be turned into foreign currency, whether in stonecrafts or as constructional materials.

Work on a number of stone projects in developing countries has shown that tremendous potential exists. In the early 1970s, the National Mineral Development Co-operation of India requested advice on the feasibility of exporting dimension stone. At the time, marble production in India was mainly for domestic consumption, and there was practically no industrialized production of granite. After identifying the potential and obtaining technological guidance, a granite industry was started. At present, India is among the five biggest granite exporters in the world.

Attention has been paid to the development of stone resources on the African continent. Quantitative figures on building stone are difficult to come by; however, it has been reported that the African share in the world market is the lowest, only 3 per cent, which comes mainly from South Africa.

Promising industrial projects are being developed in West Africa, notably grey granite in Conakry, Guinea, red granite in Aswan, marble in Ethiopia and multicoloured migmatite in Nigeria. Some North African countries which at one time were traditional suppliers are now reviving their stone industries.

An interesting application where stone, as a by-product, supplies material for low-cost housing is at Selebi-Phikwe, a nickel-copper mining town in Botswana. Blasting of the solid granite fractures the rock along fairly parallel lines. The variety of colours includes light greys, yellows, tans, browns and blacks. The rock is available from waste heaps and has been used to build rondavel-type houses which are the indigenous shelters in the area. A recent innovation has been the use of sulphur mix to cement the stones together. The sulphur comes as another by-product of many mines and can be converted relatively easily into a stabilized cement. In Selebi-Phikwe, approximately 3000 tonnes were donated for a self-help housing project.

The tuffs presently quarried around Nairobi and available in parts of Kenya have a large potential for use in low-cost housing. It is interesting that when travelling along certain roads in Nairobi, buildings on one side are built in beautiful rustic stone, while in the larger and more prestigious units on the other side of the road cement products like concrete blocks are used.

Even in countries with few stone quarries or resources there may be a thriving trade in stone. Holland has no stone deposits and yet, in 1983, besides a major stone-working machine manufacturer, there were 190 stone enterprises employing 1069 persons in the processing of imported stone.

Appendix: Stone testing

Current situation

Stone is a complex material and tests vary with both the type of stone under test and its intended end-use. Often the test methods described do not give safety limits against which to judge test results; it is left to the supplier to recommend whether the test results show that a given material is suitable for a particular application.

In the UK, testing procedures are mainly structured to allow comparison between various types of stone. The standard for kerbstones, quadrants and sets is given by BS 435 : 1975, which requires an igneous rock to be used, but no strength characteristics or testing requirements are stipulated. For tiles and slates, BS 5534 part 1 : 1978 (1985) discusses design and we are referred to other British Standards for water absorption, but again a minimum strength is not specified. The British Building Research Establishment (BRE) has outlined tests to determine saturation coefficient and porosity; these have become, as closely as is practicable, 'accepted' tests in the stone industry.

Worldwide, it is recognized that many recent failures have resulted from inadequate flexural strength and this problem is being addressed by the European Committee for Standardization (CEN). Its work is reflected in the updating of standards and specifications by both ASTM and ISO. The effects of environment and surface finish on stone properties are also being considered.

Testing procedures

Flexural strength

A stone specimen, preferably selected from the batch used for a particular project and with the same surface finish, is loaded in four-point loading, as opposed to three, and its fracture strength is measured. The test is repeated wet and dry, both parallel and perpendicular to the bedding plane. It may be repeated for differing thicknesses of specimen. The use of four-point loading means that failure occurs at the weakest point of the test piece and not directly below the applied load.

If several specimens of the same material are tested, it is possible to determine both the strength of a particular type of stone in comparison with other stone types, and the strength variations which need to be considered for that material when determining the correct safety factor. Environmental effects can be simulated by testing specimens which have been subjected to an accelerated programme of thermal cycling. The specimens are placed, face down, in a partially filled container of water which is thermally cycled between −23 and 70°C. They are tested at various stages up to a few hundred thermal cycles. It has been found that the strength decreases up to about 250–300 cycles, after which there is no further appreciable strength loss.

Crushing strength The crushing or compressive strength is tested by inserting cubes or cylinders of stone in a machine capable of exerting great pressure. The yield point is read from a dial and is calculated per unit area. The stipulated safe bearing strength should not be greater than one-tenth of the crushing strength, a safety factor of 10 (Shadmon, 1993).

Crushing strengths of stone samples

High:	2800–1800 kg/cm^2	Dense limestone
Medium:	1800– 800 kg/cm^2	Limestone, sandstone
Low:	800– 400 kg/cm^2	Porous limestone, sandstone
Very low:	≤ 400 kg/cm^2	Chalk and other powdery materials

Safety factor Stone is a natural product, and as such it exhibits large variations in strength. Environmental conditions can reduce its strength significantly. Marble, exposed to the elements, may lose 70 per cent of its flexural strength; thermally finished granite may lose 45 per cent; limestone up to 30 per cent; and polished granite up to 5 per cent. In general, an increase in absorbed moisture will reduce the compressive strength of stone; measurements of a sandstone with 15 per cent porosity showed that it lost 50 per cent of its strength when saturated.

Because of the inherently large statistical variation in the strength of stone and the effects of the environment upon it, it is necessary to apply very high safety factors when working with stone – indeed, these are unavoidable. Typically, limestone may need a factor of 8 times its measured strength; marble would require a factor of 5 in flexure and 10 in compression; granite requires a factor of 3 in flexure and 4 at fastening points.

Durability Durability has been defined (in ASTM C-119-87) as '*the measure of the ability of natural building stone to endure and to maintain its essential and distinctive characteristics of strength, resistance to decay, and appearance. Durability is based on the length of time that a stone can maintain its innate characteristics in use. This time will vary depending on the environment and use of the stone in question (for example, outdoor versus indoor use).*' This renders the much-used term 'durability' rather confusing and so it should be avoided.

To measure a material's behaviour with respect to weathering, stone samples are tested for compressive strength, specific gravity and porosity. Other similar samples are subjected to simulated attacks by frost, acid and atmospheric pollution; these are then given the same set of tests, and the change in values gives a measure of the material's environmental stability. No tests have been standardized which define a material's 'durability' when subjected to wear.

Surface finish Friction is a measure of the roughness of two surfaces pressing against each other. Static friction, when both surfaces are stationary, describes the force needed to start them moving and is a function of the total area in contact; dynamic friction describes the forces as one surface moves across another and is dependent on the speed at which the surfaces are moving. The coefficient of friction is a measure of a particular material's frictional behaviour. The aim of polishing is to obtain a very smooth reflective surface. It has been shown (Bailey and Pratt, 1955) that a smooth surface is not necessarily low in friction. This is because, when surfaces are very smooth, there is a very high contact area. A high frictional coefficient is also produced by very rough surfaces, as the unevennesses in the surfaces provide barriers to the surfaces moving across each other.

For a polished surface, a Talysurf assesses the overall roughness and a gloss meter measures its reflectivity. Both require clean, dry surfaces.

A friction meter measures the sliding friction at a given velocity, for dry or wet surfaces. The higher the frictional coefficient, the safer the floor surface.

Static friction is measured as the force required to start in motion a loaded pad of known area.

Static/dynamic coefficient measurements are useful to simulate the slippage situation. One method is to measure the resistance to continued movement of a self-powered unit across the surface being measured; the Tortus unit is one such measuring machine.

Values accepted as indicators of a floor's safety

Dynamic friction coefficient	Level of safety
0.63–1.00	Very safe
0.42–0.63	Safe
0.29–0.42	Secure
0.21–0.29	Insecure
0.00–0.21	Dangerous

Slip resistance depends not only on the initial surface finish of stones but also on their behaviour when wet or worn. For example, limestones tend to become slippery when wet, and granite loses much of its slip resistance when worn. It is therefore necessary to test polished stone surfaces under their conditions of use and to check periodically that no change in their slip resistance has occurred.

In response to increased public awareness about the need to prevent accidents caused by highly polished floors, new laws are being introduced by the EU in its Directive for Building Products and by the USA in the Disabilities Act. Nevertheless, there are currently no safety standards for slip resistance in stone, even though slipping and falling are major occupational hazards for workers in the service industries.

Bibliography

Andrew, C. *et al.* 1944. *Stone Cleaning: A guide for practitioners.* Historic Scotland and the Robert Gordon University, Aberdeen

Ashurst, J. and Dimes, F.G. 1990. *Conservation of Building and Decorative Stones.* Butterworth-Heinemann, London

Bailey, A.I. and Courtney-Pratt, J.S. 1955. 'The area of real contact and the shear strength of mono-molecular layers of a boundary lubricant'. Proceedings A227, Royal Society, London

Burton, M. (ed.) 1996. *Designing with Stone.* Ealing Publications, Maidenhead

Conti, G. (ed.) 1995. *World Stone Industry Report.* 6th edition. Societa Editrice Apuana Srl, Carrara

Gama, C.D. 1992. 'Computer simulation of waste – disposal scenario for dimension stone quarries' in Singhal *et al.* (ed.) *Environmental Issues and Waste Management.* Balkema, Rotterdam, pp. 1319–1326

Kukal, Z. *et al.* 1989. *Man and Stone.* Academia, Prague

Mackintosh, E.E. and Mozuraitis, E.J. 1982. 'Agriculture and the Agregate Industry'. Industrial Background Paper 3, Ontario Minerals and Natural Resources, Mineral Resources Branch

McLellan, A.G. 1984. 'Monitoring and Modelling: Progressive rehabilitation in aggregate mining'. *IAEG Bulletin,* Vol. 29, Paris

Montani, C. 1995. *Stone 95: World marketing handbook.* Faenza Editrice SpA, Faenza

Ordonez, S.D. *et al.* 1994. 'Physical properties and petrographic characteristics of some Bateig stone varieties'. 7th International IAEG Congress, Balkema, Rotterdam, pp. 3595–3803

Richardson, B.A. 1991. 'The durability of porous stone'. *Stone Industries,* Vol. 28, No. 10. pp. 22–25, London

Ross, K.D. and Butlin, R.N. 1989. 'The durability tests for natural building stone', BRE, Garston

Schaffer, R.J. 1972. 'The weathering of natural building stones', DSIR, Building Research, Special Report No. 18, London

Shadmon, A. 1968. 'Quarry sites surveys in relation to country planning' XXIII Int'l Geological Congress, Vol. 12, Prague

Shadmon, A. 1969. *Marble in the Philippines,* UN, New York

Shadmon, A. 1976. 'The development potential of dimension stone', United Nations, New York (ST/ESA/34)

Shadmon, A. 1979. *Stone in Nepal,* UN, New York

Shadmon, A. 1980. *Stone in Haiti,* UNIDO, Vienna

Shadmon, A. 1988. *Stone in Brazil,* UNITAR, New York

Shadmon, A. 1990. 'Dimension stone technology today', in Marinos, P.G. and Koukos, G.C. (ed.) *Engineering Geology of Ancient Sites.* Balkema, Rotterdam, pp. 1971–74

Shadmon, A. 1993. 'Dimension stone – its impact on environment and constructional applications – the role of engineering geology', *IAEG Bulletin,* No. 48, Paris, pp. 119–122

Shadmon, A. 1994. 'Environmental implications of surface treatment of building stones and ornamental rocks', 7th International IAEG Congress, Balkema, Rotterdam, pp. 3677–3682

Warland, E. 1953. *Modern Practical Masonry,* Pitman Books Ltd, London

Glossary

Abrasive finish A flat non-reflective surface finish.

Abrasive hardness Resistance to wear of stone material for floors, treads and similar uses.

Absorption Percentage of moisture by weight.

Anchor Metal device for securing stone to a structure.

Andesite A fine-grained igneous rock.

Arkose Sandstone containing 25 per cent or more of feldspars derived from granitic rock.

Arris The edge of an external angle or the salient at the intersection of two worked surfaces.

Artificial stone Man-made product that may look like natural quarried stone.

Ashlar 1. Stone with a rectangular or square face.
2. Masonry, consisting of blocks of stone, finely squared to given dimensions and laid in courses not exceeding a height of 30 cm.

Backing Part of a structure, usually the wall, to which facing units are attached or bonded.

Banker A work surface on which stone is dressed. The stone is 'banked' on the banker.

Basement 1. Mass of igneous, metamorphic and highly folded rock underlying a cover of sedimentary rock.
2. Material such as nylon gauze used to reinforce slabs.

Batter The sloping face of a wall.

Bed The plane of stratification in a sedimentary rock representing a continuous mass of sediments deposited under water at about the same one time. In igneous rock a system of natural cracks or fissures roughly parallel to each other.

Bedding planes In sedimentary or stratified rocks, the planes which separate the individual layers, beds or strata.

Bench 1. For stonemaking at mason's yard.
2. Man-made step at quarry.
3. In masonry, the lower surface upon which a block of stone rests, and the upper surface which supports the stone above.

Bleeding Staining action on stone caused by corrosive metals, oil-based putties, mastics, caulking or sealing compounds.

Breccia Rock made up of angular fragments produced by crushing due to the earth's movements and then re-cemented.

Burden The material removed from quarry face.

Buttering Placing mortar on stone units with a trowel before setting into position.

Catenary Chain cutters.

Caliche Surface deposit of carbonate by precipitation and evaporation, or capillary action.

Carve Art of shaping stone, by cutting a design to form.

Caulking Sealing a stone joint tight or leak-proof with an elastic adhesive compound.

Cavity vent An opening in joints of stone veneer to allow the passage of air and moisture from the wall cavity to the exterior.

Chamfer To cut away the edge where two surfaces meet in an external angle, leaving a bevel at the junction.

Chased A cut groove.

Clastic Consisting of rock fragments.

Cleavage Ability to break along a natural surface or parting.

Conchoidal Shell-like breakage.

Control joint Provision for the dimensional change of different parts of a structure due to shrinkage, expansion, temperature variation or other causes, so as to avoid the development of high stresses.

Coping A flat stone used as a cap on freestanding walls.

Cornice A projection which crowns a wall, a horizontal division of a wall, or an architectural feature.

Course Horizontal range of stone units along the length of a wall.

Cubic stone Fabricated dimension stone units more than 5 cm in thickness.

Cushion Resilient pad to absorb or counteract severe stresses between adjoining stone units and other materials.

Cutstone Dimension stone (originally American usage) whereas 'cutstone' originated in the Commonwealth.

Dense Term applied to sedimentary rocks with a firm, solid or dense texture and with grains firmly united or packed.

Dimension stone Stone, including marble, granites, onyx, serpentine, travertine, verde antique and others, selected, trimmed or cut to specified or indicated shapes or sizes, with or without one or more mechanically dressed surfaces.

Diorite Coarse-grained igneous rock.

Dip Degree and direction of the inclination of a bed.

Dolerite A dark-coloured igneous rock.

Dolomite Rock composed of mainly mineral dolomite.

Dowel Pin of metal inserted in holes at the edges of two facing units to connect them.

Draft A dressed strip worked on the face to the width of a draft chisel, generally for forming a true face, either straight or curved.

Edge bedding Placed at right angles to a stratum or geological bed.

Efflorescence Whitish powder, sometimes found on the surface of stones and masonry, caused by the deposition of soluble salts carried through or on to the surface by moisture.

Expansion joint Joint between stone units designed to expand or contract with temperature change.

Fabric Internal arrangement of the mineral components which indicate the aggregation process. Important modes include grains washed together, crystallized, pressed or recrystallized together.

Face Refers to the exposed surface of stone on a building or structure, or may refer to a quarry face.

Fault Fracture resulting in dislocation of bedding.

Filling Trade expression used in the fabrication of marble to indicate the filling of natural voids with cements, shellac or synthetic resins and similar materials.

Freestone A building stone workable in any direction without splitting.

Formation Group of beds with some common characteristics.

Fracture Usually, but not essentially, a break in stone.

Fraisers Surface reducers which work by milling.

Gabbro A coarse-grained, dark plutonic rock.

Gang-saw Mechanical device used to reduce marble and stone blocks to slabs of predetermined thickness, also known as frame-saw.

Ganister Type of sandstone.

Gneiss Coarse-grained rock in which bands rich in granular minerals alternate with bands in which schistose minerals predominate.

Grain The particles of crystals which comprise a rock or sediment. Also a direction of splitting in rock.

Granite A plutonic rock consisting essentially of alkali feldspar and quartz. Sometimes, commercially, all crystalline rocks of igneous origin such as true granite, diabase, dolerite, gabbro, gneiss, norite and syenite.

Greywacke Type of ill-graded sandstone, generally of dark colour, tough and indurated with a clay matrix.

Grout Mortar of pouring or pumping consistency.

Hard Hardness is resistance of substance to abrasion; in stone depends on degree of cohesion of grains and not only on actual hardness of mineral.

Header Natural break in the rock strata.

Hone finish A satin-smooth surface finish with little or no gloss.

Igneous In petrology, formed by solidification from a molten or partially molten state: one of the two great rock divisions, and contrasted with sedimentary.

Joint The space between installed stone units or between stone and the adjoining material.

Joints, jointing Fracture or parting, more or less perpendicular to rock bedding planes, along which there has been practically no displacement.

Karst Irregular rock morphology developed by the solution of surface and ground waters.

Khondalite Dark reddish-grey metamorphic rock.

Lamination Layering or bedding less than 1 cm in thickness in a sedimentary rock.

Limestone A rock consisting essentially of calcium carbonate.

Liners Structurally sound sections of stone cemented and dowelled to the back of thin stone units; to give greater strength, additional bearing surface, or to increase joint depth.

Lithology Physical character of a rock, generally based on evidence from outcrops and hand-specimens.

Masonry An assembly of building units usually laid in mortar and so arranged as to be bonded together.

Metamorphic Altered by heat or intense pressure.

Micaceous Composed of or resembling mica.

Mullion Vertical masonry detail subdividing a window.

Norite Variety of gabbro.

Onyx Crystalline form of marble.

Oolite A rock characterized by abundant small spheroidal grains 0.25 to 2.00 mm in diameter, such as oolitic limestone.

Outcrop Part of a rock stratum that appears at the surface.

Overburden Material which must be removed before mining/quarrying a useful deposit which it overlies.

Panel Single unit of fabricated stone veneer.

Parapet A wall of a building which is exposed to the atmosphere on both vertical faces and on top.

Parging Damp-proofing by applying a coat of mortar to the back of the stone.

Paver A single unit of fabricated stone used as an exterior paving material.

Peperino Volcanic rock formed by cement and volcanic sand.

Petrology The study of rocks; their origin, structure and composition.

Plutonic General term applied to the class of igneous rocks which have crystallized at a great depth and have assumed, as a rule, the granitoid texture.

Polished finish A glossy surface which brings out the full colour and character of the stone.

Pig places Shallow grooves or nicks into which pig iron is placed.

Porosity The ratio of the total volume of interstices in a rock to its overall volume. It is usually stated as a percentage.

Quarry An operation where a natural deposit of stone is extracted from the earth, through an open pit, hillside opening or underground mine.

Quarry block A rectangular piece of rough stone taken straight from the quarry, frequently dressed or wire-sawed for shipment.

Quarry sap The moisture contained in newly quarried stone.

Quartzite A rock compounded almost entirely of closely fitting quartz grains naturally cemented by secondary quartz; applied by the trade to schist.

Reaming A quarryman's term for cutting grooves on opposite sides of drill holes, to promote the straight splitting of stone, or to enlarge a bore hole.

Reinforcement Fabrication technique often called rodding refers to the strengthening of a structurally unsound marble unit. Reinforcement by lamination of fibreglass sheets to the back of the unit.

Reveal The wall surfaces at the side of an opening such as a door or window.

Rhyolite Fine grained, glassy volcanic rock.

Rift Direction of easiest splitting.

Rough-sawn A stone surface finish accomplished by the sawing process.

Rubble Stone for building purposes (walls and foundations) consisting of irregularly shaped pieces, partly trimmed or squared, generally with one split or finished face.

Sand-blasted Matt-textured stone surface finish with no gloss; accomplished by exposing the surface to a steady flow of sand or other abrasive under pressure.

Scabbling The process of removing surface irregularities from stone blocks to required size and shape for storage and shipment.

Schist A foliated metamorphic rock which splits readily.

Sealant Elastic adhesive compound used to seal stone veneer joints.

Seam A plane in which the different stone layers are easily separated.

Sedimentary rocks Rocks formed by the accumulation of sediment in water or from the air.

Serpentine A commercial marble generally dark-green in colour with markings of white, light-green or black.

Setting Installing dimension stone units.

Setting space Term referring to the distance from the finished face to the back-up material.

Sheeting planes Closely spaced joints in igneous rocks running parallel and which facilitate the extraction of sheets or slabs.

Shims Another term for 'feathers', see illustration on page 44.

Silicification Introduction of or replacement by silica.

Sill A horizontal unit of stone used at the base of an exterior opening in a structure.

Slab Piece of stone or marble cut from the quarry block after quarrying.

Slabbing Cutting or sawing into slabs.

Slate A fine-grained metamorphic rock, easily split into flat smooth plates.

Smoothing Fine grind prior to polishing.

Soundness A property of marble used to describe relative freedom from cracks, faults and similar imperfections in the untreated stone.

Spotting An adhesive contact applied to the back of a stone veneer unit to bridge the space between the unit and the back-up wall, thus helping to maintain the unit in a fixed position.

Sticking The process of cementing together broken slabs or pieces of unsound marble.

Stool A flat unit of marble, often referred to as a window sill or ledge.

Strike Direction at right angles to dip (q.v.).

Structure One of the larger features of a rock, like bedding, jointing, cleavage; also the sum total of such features, contrasted with texture.

Stylolite Suture lines in stone, mainly

limestone; known as 'crowfoot' in the Indiana limestone areas.

Syenite An igneous rock composed principally of alkali feldspar.

Tectonic Deformation by earth movements.

Texture Degree of uniformity and arrangements of constituent minerals.

Translucence The quality of transmitting light.

Travertine A variety of calcareous *tufa*, often porous.

Tufa Calcareous deposit from saturated limey water, not to be confused with tuff.

Tuffs Rocks formed from consolidated volcanic ash, generally with fragments no larger than pebble size.

Turbidite Type of sedimentary rock (often greywackes) formed from the deposit of dense slurry made up of sediment which can move rapidly down slopes.

Tympanum Space contained within a triangular or semicircular pediment.

Vein A layer, seam or narrow irregular body of mineral material, different from the surrounding formation.

Veneer A relatively thin facing of stone for buildings.

Verde antique A commercial term to describe marble composed chiefly of serpentine and capable of taking a high polish.

Waxing A trade expression used in the fabrication of interior marble to describe the process of filling natural voids.

Weep hole Opening for drainage in veneer joints or in the structural components supporting the veneer.

Wire-saw Sawing device consisting of one or more wire cables, running over pulleys, used to cut stone into blocks and slabs by diamond heads mounted on the wire or by a slurry of an abrasive and water.